Ecosystems and Human Health

Crescentia Y. Dakubo

Ecosystems and Human Health

A Critical Approach to Ecohealth Research and Practice

 Springer

Crescentia Y. Dakubo, Ph.D
Thunder Bay, Ontario, Canada
crescentia.dakubo@gmail.com

ISBN 978-1-4419-0205-4   e-ISBN 978-1-4419-0206-1
DOI 10.1007/978-1-4419-0206-1
Springer New York Dordrecht Heidelberg London

Library of Congress Control Number: 2010937571

Printed on acid-free paper

Springer is part of Springer Science+Business Media (www.springer.com)

Preface

The 2005 Millennium Ecosystem Assessment report observes that about sixty per cent of the world's ecosystem services are being degraded or used unsustainably. The report observes that over the past half a century, human activities have transformed natural ecosystems at a pace faster and extensive than in any comparable time in human history. This pace of ecosystem degradation has grave consequences for human health, including the emergence of new diseases. Since the 1970s, new diseases have been emerging at an unprecedented rate of one or more per year, with the World Health Organization confirming over 1100 epidemic events worldwide, within the past few years alone. It is anticipated that over the next few decades ecological factors will continue to play a key role in the emergence of new diseases and augment the impacts of older ones.

Since the Earth Summit in Rio in 1992, there have been increasing efforts aimed at drawing attention to the intricate interconnections and interdependencies between environment, health, and sustainable development, culminating with the recent climate change summit in Copenhagen. While these connections are being acknowledged in global and regional policy documents, their translation to influence and respond to public health and environmental problems at the lower scale still remains a challenge. For example, the health impacts of environmental degradation are experienced at the local or community level, with many public health settings struggling to contain these effects and the widespread of newer diseases. Similarly, researchers are exploring effective analytical frameworks that will provide a comprehensive understanding of the interconnections between the social, political, and natural dimensions of the environment.

These challenges and the growing emphasis on the important role of ecological factors in shaping human health, present a compelling case for rethinking current public health strategies. The intricate linkages between the social and natural components of the ecosystem require that we revisit the early 19th century's emphasis on promoting human health from a holistic and ecological perspective. While past public health research and practice sought to adopt a broader, socio-ecological view of health and to focus on broader determinants of health, the focus on individual level factors has continued to prevail, with ecological determinants receiving peripheral attention. The public health threats presented by ecological factors, now and in the future, leaves us with little choice but to refocus our efforts on identifying and developing strategies at the interface of public health and environmental management; strategies that will improve human health through the sustainable management of ecosystems.

This renewed way of thinking about improving public health has resulted in the emergence of new paradigms, such as the *ecosystem approach to human health*, or the *ecohealth approach*, the subject of this book. The ecosystem approach to human health bridges thinking in the public health and the natural resources management fields, and explores ways to understand and manage the various components of the ecosystem so as to improve human health and well-being. The ecohealth approach seeks to promote a holistic view of health, with environmental sustainability as a major component of this overall well-being. The ecohealth approach encourages research, practice and policy that aim to improve human health and well-being through better ecosystem management interventions. The emphasis on both human health and ecosystem health underscores the interdependencies between the two systems, and provides a means for achieving broader goals of sustainable development.

From a research perspective, the ecohealth approach integrates indigenous perspectives with the views of experts from the natural, social, and health sciences, to investigate and respond to problems at the interface of environment and health. The approach makes use of a transdisciplinary team of researchers who engage relevant stakeholders and beneficiaries of the problem under investigation in all aspects of the research process. Participatory research procedures are central to the ecohealth approach, with the ultimate goal of generating increased understanding of the causal basis of ecologically-mediated health problems, and to raise people's consciousness to respond to their health concerns in a proactive manner.

Since the 1990s, the concept of an ecosystem approach to human health has been gaining widespread attention. A number of institutions around the world have begun to adopt the ecosystem approach to promote public health. For example, in Canada, the International Development Research Centre (IDRC) can be referred to as a pioneer in spearheading the application of this approach in developing countries, and in developing Communities of Practice around the world. Some medical schools have also begun to incorporate ecohealth concepts and principles into their curricula.

However, as an emerging field, ecohealth lacks the theoretical rigor that is often seen in other public health sub-disciplines such as medical sociology, health geography, and medical anthropology. Most often than not, the conventional ecohealth literature adopts concepts and notions of "health", "ecosystem degradation", and "community participation" without re-evaluating how these are constructed, and how social and political framings are woven into these constructions. Also, the causal basis of ecosystem degradation tends to be attributed to factors such as rapid population growth, "inappropriate" land use practices, and poverty, without considering how these factors have been shaped by unequal power relations that characterize human-environment relationships and represent coping strategies and forms of resistance. Similarly, ecosystem-mediated health problems tend to be attributed to "inappropriate" interactions with the biophysical environment and consequently exposure to disease vectors and pathogens, while failing to take into account the socio-political factors that caused the disease-prone environment in the first place. In addition, it is important that attention be paid to how we construct subject positions such as the "sick" and "healthy". These constructions need to be evaluated through the lens of how ecohealth is deployed as a discourse, and taking care to ensure that ecohealth knowledge claims are transparent.

This book is designed to take ecohealth research and practice to this next level, the adoption of a critical lens. The book draws on critical social theory to examine public health and environmental problems. In particular, it draws on theoretical perspectives from political ecology (of health), the sociology of science, poststructuralism, postcolonial, and feminist theories as applied in public health and environmental discourses. Building on these, the book lays the contours for a new framework – *A Critical Approach to Ecohealth Research and Practice*, which bridges thinking in critical public health and critical political ecology.

In addition to proposing a critical lens to ecohealth research and practice, the book walks students, researchers, and practitioners through the practical processes of conducting an ecohealth research project, from gaining entry into the research site or community, to conducting a culturally and socio-politically conscious research project.

The case studies presented in this book draw on my experience as an ecohealth research practitioner and explore the methodological and ethical challenges mostly encountered when embarking on a community-based ecohealth research project. The application of the ecohealth approach to Indigenous health concerns is also explored, as well as an examination of on-going efforts by global and regional initiatives to integrate environment and health policy and to link this with broader public policies.

For purposes of organization, this book has been divided into four parts. Part I – Ecohealth: The Ecosystem Approach to Human Health, which includes Chapters 1, 2, and 3, and reviews the literature on the linkages between health and environment and traces the events in both the public health and environmental fields that led to the re-emergence of the ecosystem approaches to public health. It also describes the key concepts and principles of the ecosystem approaches to human health. Part II – Methodological Approaches and Processes to Conducting Ecohealth Research, which includes Chapters 4 and, 5 describes the methodological approaches and processes for conducting ecohealth research. It outlines the key elements and principles of community-based participatory action research and a healthy community strategic planning process. In addition, it describes a step-by-step, practical approach to conducting an ecohealth research, from forming a transdisciplinary research team to collaboratively analyzing and implementing the research findings. Part III – Case Studies: Application of the Ecohealth Approach, which includes Chapters 6, 7, 8, 9, and, 10 examines the application of the eco-health approach to investigating environment and health concerns. Chapters 6 and 7 present the findings of an ecohealth project that was conducted in a West African community, and discusses how political ecology and community strategic planning processes were used to help a community investigate and respond to its environment and health challenges, and also plan for a healthy community. Drawing on the case studies presented, Chapter 8 explores some of the methodological and ethical challenges encountered when conducting an ecohealth project. Chapter 9 explores the application of the ecohealth approach to Indigenous environmental health concerns. Chapter 10 examines how efforts are being made regionally and globally to develop integrated health and environment policy frameworks and to translate these to inform programming at the local level. Part IV – A Critical Approach to Ecohealth Research and Practice, which includes Chapters 11, 12, 13, and, 14 lays the foundation for a critical approach to ecohealth research and practice. It explores the key elements of critical social theory, examines how these are applied to environment and public health issues, and then articulates a critical framework for ecohealth research and practice.

As one of the first books to introduce the concept of a critical approach to ecohealth research and practice, the goal of this book, then, is to not only introduce students, researchers, and practitioners to the ecosystems approach to human health, but also to stimulate critical thinking and the application of critical theoretical perspectives to examining the complexities surrounding people, environment and health relationships. It is through the adoption of such a critical lens, that we are able to produce knowledge claims that are socially and ecologically relevant, as well as develop interventions that are liberating and not constraining.

This book could not have come to fruition without the support of my family, friends, colleagues and mentors. In particular, I am grateful to the examining board and academic committee of my doctoral dissertation, who encouraged me to put down, in the form of a book, my transdisciplinary training and experience and to stimulate this new line of thought in the emerging field of ecohealth. My heartfelt gratitude goes to my family – Collins, Ethan, Bernard, Zaneta, and my dedicated husband Gabriel, for their continuous support and encouragement.

Thunder Bay, ON, Canada   Crescentia Y. Dakubo

Contents

Part I
Ecohealth: The Ecosystem Approach
to Human Health

Chapter 1
Exploring the Linkages Between Ecosystems and Human Health

Contents

1.1 Introduction

The linkages between human health and ecosystems are complex, dynamic, and political. For millennia ecosystems have provided humans with essential services such as food, water, shelter and medicine. At the same time, they have mediated the transmission of many diseases and posed a number of health risks. The vitality of ecosystem services for human health and well-being is well captured by Bernard Abraham, President of Weskit-Chi Aboriginal Trappers Association, when he commented on the importance of forest ecosystems to Aboriginal people. He observed that many Aboriginal people consider the forest as: "their food bank, drugstore, meat market, bakery, fruit and vegetable stand, building material centre, beverage supply, and the habitat for all of the creator's creatures."[1] Many Indigenous people across the world consider the health of the "country" to be intricately linked to human health and community health and well-being. This sentiment is not only

---

[1] A quote by Bernard Abraham, President of Weskit-Chi Aboriginal Trappers Association. http://www.envirowatch.org/gndvst.htm. Accessed May 01, 2010.

C.Y. Dakubo, *Ecosystems and Human Health*, DOI 10.1007/978-1-4419-0206-1_1,
© Springer Science+Business Media, LLC 2011

3

true for Indigenous people, but for society as a whole. In addition, the intricate links between ecosystems and human health is expressed through aspects of the Indigenous culture, including views and notions of holistic health and well-being, and ecosystem-based cultural rites, and overall close proximity to nature. The World Health Organization captures this notion of holistic well-being when it defines health as *a state of complete physical, mental, and social well-being and not merely the absence of disease or infirmity* (World Health Organization 1948). However, it is the Ottawa Charter for Health Promotion that makes explicit the connection between human health and ecosystem health through its identification of "stable ecosystems" and the "sustainable use of natural resources" as essential components for health improvement (WHO 1986).

Despite this close association between human health and ecosystem health, recent evidence from the Millennium Ecosystem Assessment (MA 2005) suggests that global ecosystems are failing in their ability to continue to provide the services that are essential for human health and well-being, because of increasing human pressure on ecosystems worldwide. The report indicates that, over the past half century, human activities have altered the natural ecosystem more rapidly and extensively than in any comparable time in history. This increased pressure has resulted in close to 60 per cent of the world's ecosystem services being degraded or used unsustainably (ibid). This trend will likely continue to the extent that human beings continue to depend on ecosystems for both the necessities of life such as food and water, as well as the luxuries of diamond and caviar. What is required is recognition of the conjoint nature of society and ecosystems and how each cannot be managed separately. Human beings are integral to ecosystems, and for sometime now, the environmental sector has made use of integrated approaches to ecosystem management that have tried to balance community, economic, and environmental needs. However, the question that still exists is to what extent are these balanced? In addition, what has eluded many concerned with understanding and responding to the underlying causes of human-induced ecosystem degradation is their political and illusive nature. How we apportion blame to the causal factors responsible for ecosystem degradation, and poor health, must be subjected to rigourous analytical interpretations.

The factors that have commonly been identified as responsible for ecosystem degradation focus on rapid population growth, abject poverty and poor land use practices in the global South, and over consumerism in the North. However, as has been demonstrated throughout this book, such factors are the outcome of underlying problems that are not readily apparent, and hence are insufficient to explain the causal basis of environmental problems. The causal basis of ecosystem degradation must be examined from the basis of the structural inequalities surrounding the use of and control of ecosystem services, and how such inequalities drive various land use practices that degrade ecosystems. For example, it is important to examine how states, corporate giants, mining and logging companies, and Northern interests monopolize the environments of weaker actors (e.g. local farmers, peasants, global South), forcing them to till and live on marginal lands. In an attempt to eke out a living, displaced or marginalized communities encroach on fragile and protected lands to "illegally" access natural resources, sometimes making use of "inappropriate" land use practices (Bryant and Bailey 1997; Bryant 1998). Also, in an attempt to increase productivity and maximize the potential of marginal lands, weaker actors may resort to intensive production systems, making use of fertilizers, pesticides, and intensive farming practices which not only adversely impact ecosystems, but consequently human health. In such contexts, then, the identification of "inappropriate land use practices" as one driving factor of ecosystem degradation completely misses the underlying reason for the use of inappropriate land use practices, and any attempt by "experts" or professionals to develop interventions or policies to correct this will be ineffective and may not prevent/correct ecosystem degradation. The causal basis of human-induced ecosystem degradation is complex and not readily apparent to external scientific experts. It is important that any perceived causal factor be examined from an historical, socio-political, and economic perspective, and examined through the lens of unequal power relations surrounding human-environment relationships. In addition, ecosystem degradation should be linked to broader processes such as unfair trade agreements, global prioritization of environmental problems and the desire for the purest gems, chocolate, coffee and caviars have contributed to destroying vast ecosystems, especially in tropical and developing regions of the world. The unfair trade agreements between the global South and the North, the re-colonization of Africa by China, and the lax environmental regulatory environment in poorer regions have all resulted in a monopoly of Southern ecosystems by multinational companies, resulting in the use of unsustainable methods of resource extraction and ecosystem degradation.

One other dimension of examining the causal explanations of environmental problems that is of interest to critical scholars is how environmental problems such as ecosystem degradation comes to be identified, defined, and labeled. How do we come to understand ecosystems as "extensively degraded", or describe other environmental problems as "global crisis"? Critical scholars argue that the identification and causal explanation of environmental problems is not value-free nor is it ever politically neutral. They argue that the framings of environmental problems and their causal explanations are shaped by the social and political contexts within which they emerge, and so are never partial, but can be located. Critical environmental scholars caution against accepting scientific environmental knowledge claims as "fact", "accurate" and a true representation of reality, without re-evaluating these claims within the socio-political and historical contexts within which they are framed (Bryant 1998; Forsyth 2003; Peet and Watts 1996).

With the exception of phenomena such as climate change, although this has come under scrutiny recently, critical scholars are concerned that by representing many environmental problems as "global" in scope and "crisis" in nature, we gloss over the particularities of environmental problems in specific localities, and fail to pay particular attention to the varying experiences and coping abilities of different population groups. We also fail to capture the micro-politics and power struggles surrounding access to, and use of natural resources at varying scales, and how such struggles shape people's interactions and experiences with the biophysical environment. Also, given that most scientific environmental knowledge serves as the basis for policy formulation, there are concerns that environmental policies and interventions that are based on "unreconstructed" scientific knowledge could fail to uncover the "real" causes of ecosystem degradation and end up proposing wrong interventions that could further augment existing inequities surrounding the use and control of those resources (Forsyth 2003). For example, in many parts of Africa, some natural resource management policies are still based on colonial policies, without reevaluating these policies within the current challenges and needs of today's communities.

Within the circles of public health, similar critical perspectives have been used to interrogate public health knowledge claims, the constructions and explanations of certain public health problems, and how subject positions are constructed and labelled. Just like critical environmental scholars, critical public health scholars seek to illustrate that the emergence and social patterning of specific public health problems, especially those associated directly and indirectly with the environment, lie in the unequal power relations surrounding the use of, and control of environmental resources and the uneven distribution of the associated costs (e.g. pollution) of environmental activities. The concern here is to broaden the causal basis of the emergence of specific ecosystem-mediated health problems beyond exposure to disease vectors and microbes, to incorporating people-environment power dynamics and how such dynamics result in the exposure of weaker actors to environmental health risks. Explaining health problems from the basis of exposure to microbes and pathogens alone is equivalent to blaming the victim, while relieving important social and political factors that constrain the individual from freely making the decisions to avoid risk in the first place. Similarly, the uneven distribution of the costs of ecological activities and the resulting social patterning of poor health should be examined from the context of how unequal power relations allow powerful actors to displace their environmental costs to weaker actors through such acts as dumping toxic waste in other communities. These communities also are those with limited coping capabilities and have little resources to mitigate the adverse effects of these environmental pollutants.

The above concerns illustrate the complex dynamics surrounding people-environment relationships, illustrating that such issues cannot be understood through uni-lateral analytical procedures, but instead must be contextualized to reflect the temporal and spatial dimensions of such phenomena. Ecosystem degradation, the causal explanations, and the associated health problems are equally complex and should not be explained simplistically and uni-dimensionally. After all, ecosystems and society are conjoined and the activities within each sphere must be seen as constituted by, and from the other. Such inter-dependencies (political, social, economic, ecological dimensions) must always constitute the core of ecosystem-society-health investigations. The theoretical frameworks used to analyze human-environment interactions from such political, ecological, and social perspective fall outside the purview of pure ecological and health sciences disciplines, instead residing more with transdisciplinary fields such as health geography, environmental sociology, medical anthropology, and other related fields. Similarly critical perspectives have been applied to environmental issues in the form of post-structuralist or critical political ecology and to public health field in the form of critical public health. However, the extent to which the two fields have been brought together to examine issues at the interface of public health and environmental conditions is very limited.

It is the goal of this book to draw on a variety of critical theoretical perspectives from the social, natural and health sciences to develop a rigorous theoretical framework that will allow for a critical examination of problems at the interface of environment and health, or simply referred to as ecohealth concerns. Most of the ecohealth literature has not engaged with such theoretical perspectives and hence lacks critical theoretical rigour in its analyses of environment and health phenomena. This book draws on critical social theory, including political ecology, feminist theories, and postcolonial and poststructuralist perspectives to examine environment and health issues from a critical perspective. In so doing, a new analytical framework called *critical ecohealth* is developed through the fusion of two theoretical perspectives: *critical political ecology* and *critical public health*. Critical ecohealth locates ecohealth problems, their causal explanations, the proposed interventions within a broader analytical framework, examining how they are framed, and drawing attention to their socio-political, economic, and historical antecedents. Prior to examining these theoretical frameworks in subsequent chapters, it is important to review some of the common associations between human activities, ecosystem change and how this influences human health and well-being.

1.2 Ecosystem Services and Human Health

The Millennium Ecosystem Assessment report (MA 2005) describes an ecosystem as *a dynamic complex of plant, animal, and micro-organism communities and the nonliving environment interacting as a functional unit.* Ecosystems including, farmlands, water bodies, woodlands, rangelands, and forests, produce services that are essential for human health and well-being. These services are usually classified into four categories: provisioning services, regulating services, cultural services, and supporting services (MA 2005). *Provisioning services* refer to the benefits derived from ecosystems such as food, freshwater, fibre, shelter, medicine, and fuel. These basic necessities underlie the sustenance of many communities. Many developing country nationals rely on the natural environment for medicinal plants, wildlife and other non-timber forest products. The second category, *regulating services*, refer to ability of ecosystems to regulate climate, purify freshwater, and regulate pest and diseases. The regulation of ecosystem processes can modify ecosystems in ways that influence the proliferation and transmission of disease vectors, such as mosquitoes or snails. *Cultural services* are those non-material benefits obtained from ecosystems, and include aesthetic, spiritual, educational, and recreational qualities. Cultural services provide a wide spectrum of benefits, since different cultures associate and interact with the environment in myriad ways. For example, for some, the natural landscape serves as an ideal space for healing, meditation, recreation and the performance of cultural rites. The close bond between many Indigenous communities and other local people, and the biophysical environment endows them with unique knowledge systems about the structure and functioning of these ecosystems. This local knowledge is an essential complement to scientific understanding of the environment. Lastly, *supporting services* are those that contribute to ecosystem processes such as primary production, soil formation, and nutrient recycling. Compared to other services, the benefits of supporting services are indirect and occur over a longer-time frame.

The growing need to meet societal demand for food, shelter, livelihood and profits has resulted in increased pressure on ecosystems, and compromising their ability to continue to provide ecosystem services at an optimal level. Land use activities such as deforestation, clearance of virgin lands for agriculture and human settlement, irrigation, dam construction, road building, mining, wetland modification, and urbanization, have been identified as some of the key modifiers of ecosystems around the world. The modification of various ecosystems has resulted in the emergence and spread of a number of infectious diseases, and modified the transmission of endemic diseases (Patz et al. 2000). For example, the clearance of forests for agricultural purposes can disrupt the structure and functioning of ecosystems and lead to the emergence of infectious diseases. In Central Africa, the outbreak of Ebola, which killed hundreds of people and thousands of apes was linked to human migration into forested ecosystems where people came into contact with new microbes and animal reservoirs (Leroy et al. 2004). In Malaysia, agricultural activities have been linked with the emergence of Nipah virus (Lam and Chua 2002), while increased risk to Lyme disease in the northeastern United States has been associated with forest fragmentation, biodiversity loss, followed by suburban housing development (Schmidt and Ostfeld 2001). Also, with the relative ease of travel and transportation of goods and services around the globe, it does not take long for infectious diseases to spread from one corner of the globe to the other, as seen with the recent case of Severe Acute Respiratory Syndrome (SARS) and Swine Flu.

Infectious diseases continue to be of grave concern, both in the developed and developing world. Within the past few years alone, the world has seen the emergence of new infectious diseases such as SARS, H1N1, and HIV/AIDS. Not only did the emergence and quick spread of these diseases cause global pandemonium and drain the health budgets of many regions, but also has raised concerns about the state of global public health security and the readiness of public health authorities to respond and contain the spread of these pandemics in a quick and effective manner. Factors such as rural-urban migration, globalization, North-South migration, trade, and the fast pace of travel all contribute to making this task a challenging, yet important one. According to a recent report, not only are infectious diseases spreading faster geographically than any period in history, but they also seem to be emerging at a quicker pace than before (WHO 2007). The report indicates that since

1970, new diseases have been emerging at an unprecedented rate of one or more per year, resulting in about 40 new diseases that were unknown about a generation ago. Also, within a 5-year period leading up to 2007, the World Health Organization had confirmed over 1100 epidemics events worldwide.

This trend does not seem to be abating any time soon and re-iterates the need for a comprehensive understanding of the mechanisms through which human-induced ecosystem change adversely impacts human health. It is important to understand the clear linkages between our activities, what drives these activities, how these activities transform the ecosystem into a disease environment or a health-promoting environment, and how we are differentially impacted by such transformations. Such understanding must be informed by the social, political, and historical contexts within which ecosystem change occurs, so as to allow for the development of socially and biophysically relevant interventions. The sections below explore the linkages between some land use activities and how they shape human health outcomes.

1.3 Land Fragmentation and Health

Activities such as deforestation, clearance of virgin lands for agricultural purposes and human settlement, and road construction for mining and logging are some activities that have led to increased fragmentation of many terrestrial ecosystems. These land use activities disturb ecosystem balance and pre-existing conditions that serve to modulate the emergence and interaction of disease pathogens. This disturbed equilibrium brings humans into contact with new pathogens that can infect humans, livestock or wildlife (Wolfe et al. 2000). The emergence and re-emergence of many infectious diseases such as Chagas disease, trypanosomiasis, leishmaniasis, and onchocerciasis has been associated with land use changes (Molyneux 1998). Habitat changes also favor the emergence of zoonotic diseases and many mosquitoe-borne diseases (Gubler 2002).

Recently, there have been increasing concern about the public health threat posed by zoonotic diseases. Zoonotic pathogens – that is, those pathogens that can be transmitted between wild or domesticated animals and humans – have been identified as the most significant cause of emerging infectious diseases (Taylor et al. 2001). Taylor and colleagues observed that out of 1415 species of infectious organisms that have been identified to be pathogenic to human beings, 61% are zoonotic pathogens. Emerging infectious diseases such as Severe Acute Respiratory Symptoms (SARS), avian influenza, West Nile and HIV/AIDS, Nipah virus, Ebola, and hantavirus pulmonary syndrome are all associated with zoonotic pathogens. In general, zoonotic diseases are usually severe, with high fatality rates, and have no readily available cure, treatment or vaccine. Because zoonotic pathogens complete part of their natural life cycle in animal hosts, any human-induced activity that disturbs the equilibrium of wildlife habitats, such as encroachment into forested areas, is likely to facilitate the transmission of zoonotic pathogens between humans, wildlife, domestic animals, and plants (Daszak et al. 2001).

Land use activities such as tropical deforestation and the processes leading to it have also been associated with the emergence and proliferation of diseases such as malaria, especially in Africa, Asia, and Latin America (Coluzzi 1994; Tadei et al. 1998). The clearance of forested areas for agriculture, rangelands and settlement allows people to inhabit previously uninhabited spaces, thus exposing them to new disease pathogens (Kalliola and Flores Paitán 1998). The construction of forest roads, the creation of culverts and other dugouts collect rainwater and serve as breeding grounds for mosquitoes (Patz et al. 2004). Also mercury is naturally embedded in the soils of most rainforests. Hence soil erosion that occurs following downpours wash mercury residue into rivers and other water bodies, contaminating water bodies. Such scenarios have led to contaminated fish in places like the Amazon (Fostier et al. 2000).

Another example of the health implications of deforestation is noteworthy. In northeastern United States, partial deforestation, followed by subsequent land use changes and human settlement patterns led to the emergence of Lyme disease (Glass et al. 1995). Lyme disease is a bacterial disease that is transmitted by the bite of a deer tick. Rodents are the major reservoir hosts for the bacteria, while deer serves as the host for the tick vectors (Steere et al. 2004). Lyme disease was first named in 1977, but discovered earlier. Incidence has been reported in North America, Asia, and Europe (ibid).

Finally, in addition to logging, mining is one extractive activity that causes a number of health problems. In many regions in Africa, lax environmental regulations prevent mining companies from taking the necessary steps to ensure their activities cause minimal impacts to both human and ecosystem health. In tropical rainforests, the use of mercury to extract gold from riverbeds has contaminated fish in many rivers, rendering them toxic (Lebel et al. 1998). Also the land degrading activities associated with mining has caused some communities to lose farmlands and livelihood options. Dugouts, culverts and mining pits create favourable breeding grounds for mosquitoes.

1.4 Water Resource Development and Health

Human interventions in watersheds, rivers, and lake systems take many forms including: irrigation, aquaculture, river damming and other watershed activities. Most of these activities interfere with the natural functioning of aquatic ecosystems, and may inhibit their ability to provide ecosystem services, such as regulation of the hydrological cycle and filtration of freshwater. Some of these activities also alter watersheds in ways that create conducive environments for the proliferation and transmission of disease vectors such as snails and mosquitoes. Some commonly identified diseases emerging from human-induced transformation of watersheds include malaria, dengue and Japanese encephalitis, shistosomiasis, onchocerciasis, and trypanosomiasis.

Crop irrigation and the construction of dams are two land use activities that alter aquatic habitats and affect the proliferation, survival, and distribution of disease vectors. For example, irrigated rice fields provide good breeding grounds for mosquitoes, and have resulted in increased incidence and transmission of malaria in Africa, and Japanese encephalitis in Asia (Keiser et al. 2005). Also, culverts, ditches, canals and ponds associated with irrigation provide ideal conditions for the proliferation of mosquito species such as the *Culex tarsalis*. *Culex tarsalis* is a mosquito species that bites both humans and animals, and as such a major bridge vector for diseases that are constantly present in animal populations, such as the encephalitis that occurred in St. Louis in western United States (Mahmood et al. 2004). Also, irrigation activities in the Nile Delta following the construction of the Aswan High Dam resulted in the proliferation of another mosquito species, the *Culex pipiens,* which is associated with increased soil moisture levels. The *Culex pipiens* is associated with the arthropod-borne disease Bancroftian filariasis or elephantiasis, which mostly occurs in Africa and other tropical regions (Harb et al. 1993; Thompson et al. 1996).

Microbial contamination of water as a result of inappropriate sanitation and hygiene is still pertinent, especially in developing countries. A recent report from the World Health Organization estimates the burden of disease from inadequate water, sanitation and hygiene to amount to 1.7 million deaths annually, with over 54 million healthy life years lost. Also, water-associated infectious diseases claim up to 3.2 million lives each year, approximately 6% of all deaths globally (Prüss-Üstün and Corvalán 2006). The contamination of drinking water sources is not only pertinent to the developing world, but also the developed. For example, intensive farming practices and poor food processing in industrialized countries can lead to the contamination of public water sources, as was seen in the Walkerton case in Canada. In 2000, Canada experienced its worst ever water contamination, when a small town in Ontario, Walkerton, got its public water supply infested with *Escherichia coli* *(E. coli)* bacteria from farm runoff. The incident resulted in the death of seven people, with as many as 2,300 falling sick.[2]

Aquatic ecosystems serve as natural reservoirs for the cholera bacterium (*vibrio cholerae O1*), where it remains dormant in phytoplankton and zooplankton (Colwell 1996). Environmental conditions that cause algal blooms, such as climate-induced warming of waterways and eutrophication by agriculture and domestic nitrate and phosphate runoff, may increase the proliferation of zooplankton leading to increased dissemination of cholera into human populations (Levins et al. 1994). Also, there is increasing evidence suggesting that the seasonality of cholera epidemics may be linked to the seasonality of algal blooms, and the food chain in marine ecosystems (Colwell 1996). It is recommended that monitoring algae and other microscopic marine organisms for vibrio, especially using remote sensing satellites, may help

---

[2]http://www.cbc.ca/news/background/walkerton/ Accessed May, 10th, 2010.

establish an early warning system for detecting emergence of the pathogen (Levins et al. 1994).

Water bodies that are contaminated through the use of pesticides and other toxic chemicals can also pose serious health risks to people, and adversely affect various organ systems. For example, exposure to low concentrations of chemicals such as PCBs, dioxins and DDT may interfere with normal hormone-mediated physiology, impair reproduction, or cause endocrine disruption (Prüss-Üstün and Corvalán 2006).

1.5 Urbanization and Health

On April 7th each year, the World Health Organization (WHO) celebrates World Health Day. It selects a key global health issue as the theme for the day and generates awareness of the problem globally, nationally and locally. For 2010, the theme for World Health Day was "Urbanization and Health". The WHO identifies urbanization as one of the biggest health challenges of the twenty-first century (World Health Report 2008). This is based on the realization that urbanization is proceeding faster than cities can build the necessary infrastructure to contain the increasing numbers. In 2007, for the first time in history, the world's urban population surpassed 50%, with the projection that this number could exceed 70% by 2050 (UN-Habitat 2006). By 2030, it is expected that six out of every 10 people will be living in the city, and by 2050 this figure is expected to increase to 7 out of every 10 people (ibid).

Rapid and unplanned urbanization has numerous health implications, not just for the urban poor, but also for all city dwellers. It is true that the urban poor will bear the disproportionate burden of urban health problems. However, lack of social services, employment opportunities, education and other services engender despair, violence, and increased vulnerability. These problems are usually not constrained to only urban slums, but permeate to the suburbs and affect the entire society. It is therefore important that urbanization health-related concerns be viewed from a broad perspective, and their solutions be incorporated into broader public policies. The public health challenges facing urban ecosystems span beyond the health sector and must addressed from an integrated and intersectoral perspective, with partnerships among all relevant sectors.

In addition, it is important not to lose sight of the health conditions specific to urban slums. Currently, over 1 billion people – about one third of the urban population – live in slums, with this figure is expected to increase to 1.4 billion by 2020 (UN-Habitat 2006). Inequitable access to most social services, poor housing and sanitation, and inadequate water supply characterize the conditions in many urban slums. These conditions make urban slums fertile grounds for the proliferation and transmission of communicable diseases. Common health problems of the urban poor include tuberculosis, HIV/AIDS, and chronic diseases such as diabetes, heart disease, mental disorders, and road traffic accidents, and drug-related deaths.

With such clear trends of increasing urbanization, perhaps what is required is to refocus efforts on preventive health, improving living conditions in urban centres by investing in infrastructure for sanitation, water supply, and supportive housing. It is also important for public health authorities to prepare for the onslaught and complex urban health problems that could arise with such increasing trend. Also, urban planning must ensure that urban centres become welcoming and inclusive communities with all the necessary amenities to cater to the wide spectrum of cultural diversity that immigrates to urban areas. Without such readiness, urban health problems could be a time bomb waiting to explode with the onset of any pandemic.

Finally, meeting the needs and wants of city dwellers takes a great toll on suburban and rural ecosystems, which leaves behind bigger ecological footprints. Similarly, the demands in the North for coffee, cocoa, burgers, quality furniture, and minerals take a toll on Southern peripheral ecosystems. The extraction and processing of resources such as timber and minerals fragments ecosystems and increases the opportunity for the emergence of new diseases. Also, aquaculture, shrimp farming, and deforestation for agriculture and ranching all destroy ecosystems. In most cases, the ecosystems drawn on to satisfy the needs of urban dwellers are usually not within the immediate vicinity but in remote, rural areas or in developing countries and tropical regions. In this case, the immediate and direct impacts of ecosystem destruction are displaced to inhabitants of these ecosystems, not the city dwellers. Due to their poor status and limited resources, these communities are unable to cope with or take adequate steps to mitigate the adverse impacts of ecosystem destruction on human health.

1.6 Modern Food Production Systems and Health

The increasing demand for livestock products, especially pigs and chickens, has led to the use of intensive, industrial, and landless production systems (Delgado et al. 1999). These intensive production systems, in association with ecological and other factors, have been linked to the emergence of diseases such as bovine spongiform encephalopathy (BSE), Severe Acute Respiratory Syndrome (SARS), Nipah virus, and avian influenza (World Health Report 2007). Modern production systems are characterized by activities such as increased livestock trade between regions, especially poultry and wild animals (bushmeat), overcrowding and mixing of livestock breeds, and cohabitation of livestock and people, especially in rural communities (Graham et al. 2008). Such production practices create fertile grounds for interspecies host transfer of disease agents, resulting in the emergence of novel strains of diseases or human pathogens such as SARS and influenza.

While modern production and processing systems have led to increased availability of food and livestock products, they have also increased pressures on ecosystems: fragmenting habitats, polluting environments and posing serious human health risks. For example, intensive production systems usually require large quantities of livestock feed and increase the pressure on cultivated ecosystems. They also make use of large quantities of fertilizers, pesticides and water to enhance productivity. Intensive farming practices also generate large amounts of waste, which sometimes is not adequately disposed off. Waste is mostly flushed into waterways, which end up polluting freshwater bodies, contaminating public water supplies, and affecting marine plants and animals. In addition, some intensive livestock management practices routinely use sub-therapeutic antibiotics, which have resulted in the occasional emergence of antibiotic-resistant strains such as Salmonella, Campylobacter and *E. coli* bacteria (Garofalo et al. 2007).

The recent outbreak of Nipah virus in Malaysia is a typical example of a disease that occurred as a result of animal husbandry in association with other factors. Nipah virus is an emerging viral pathogen that causes encephalitis, an inflammation of the brain. It is fatal in up to 75% of the people it infects (WHO 2007). Between 1998 and 1999, the first outbreak of Nipah virus was reported in the Malaysian Peninsular, where 265 human cases including 105 deaths were reported (FAO/WHO 2002). The emergence of the virus is attributed to the interaction of various factors including expanding human population, climate change, poor governance, illegal land clearing, forest fires and intensive animal husbandry (ibid). The path of contagion is traced back to the human cases coming into direct contact with sick or dying pigs or fresh pig products. These pigs became sick after coming into contact with a flock of bats that were infected with a previously unknown virus. The bats migrated from neigbouring Indonesia, following an intense El Niño dry spell and forest fires in the region. In Malaysia, the bats came into contact with intensively, commercially raised pigs that were located near fruit orchards. The pigs acted as the intermediate hosts of the new virus, and developed respiratory illnesses. It is believed that transmission among pigs occurred through the aerosol route, with transmission from pigs to humans taking place following contact with throat or nasal secretions of pigs by humans. Nipah virus later occurred in Singapore, where it infected 11 human cases resulting in one death. In Malaysia, the outbreak ended with the mass culling of more than 1 million pigs (WHO 2007).

Recent findings suggest that the virus may have become more pathogenic for humans following the outbreaks in Malaysia and Singapore. This means that the virus can spread to humans without an intermediate host such as the pig, and the transmission from human to human can occur with casual contact. For example, evidence suggests that, in the most recent outbreaks in Bangladesh and India, the consumption of contaminated food such as fruits contaminated with the urine or saliva of fruit bats could likely constitute the route of exposure for several new human infections. Also human-to-human transmission could occur through close contact with people's secretions and excretions. In Siliguri, India, it was observed that transmission of Nipah virus occurred in health care setting, where close to 75% of the cases occurred among hospital staff and visitors (WHO Fact Sheet on Nipah Virus 2009).

1.7 Climate Change and Health

Leading up to the United Nations Summit on climate change in Copenhagen, there have been a number of discussions and media coverage on the potential health effects of climate change. For example, the journal *Lancet* dedicated an entire series to *climate change and health*. The increased attention and wide coverage of climate change generated both awareness and skepticism, leading some to question the accuracy of climate change data, and to assess whether or not climate change has become one of those phenomena, whose scientific explanations and claims are politically and self-interest driven. While, this dialogue is on-going, there are also discussions about the health and potential health implications of climate change.

Climate change is expected to have both direct and indirect impacts on human health and well-being. Extreme weather events, sea level rises, and temperature changes are expected to adversely impact ecosystems, and inhibit their ability to continue to provide the essential services needed for good health, including the provision of clean air, safe drinking water, adequate food supply, shelter, and medicinal plants. Ecosystems play a vital role in regulating climatic conditions through cooling and warming mechanisms, preserving the balance among species, and acting as sinks for greenhouse gases and other pollutants. Climate-induced changes will likely disrupt the ability of ecosystems to continue to fulfill these functions.

For the most part, climate change is expected to increase the incidence and impacts of some of the world's leading killer diseases, such as malaria, diarrhoea, dengue, and malnutrition. These health problems and the pathways leading to their occurrence are highly sensitivity to climatic conditions. For example, climate-induced change can affect the proliferation, distribution and transmission of disease vectors and can also influence the length of transmission seasons for vector-borne diseases. Extreme weather events, such as floods and windstorms may contaminate fresh water supplies, facilitate the dispersal of microbes, and affect the breeding, survival, and abundance of disease vectors. Outbreaks of diseases such as cholera and leptospirosis have followed flooding in Central America (Wilson 2000). Heavy precipitation may pollute water sources with increased quantities of chemical and biological pollutants that are washed into rivers and from overloading sewers and waste storage facilities. Temperature increases may also affect water quality by increasing the growth of microorganisms and decrease dissolved oxygen (McMichael et al. 2006).

Temperature-related impacts are varied. Rising temperatures may cause drought, increase demand for irrigation, and negatively affect crop production, leading to increased malnutrition, especially in developing countries (McMicheal 1997). Changes in temperature and humidity may affect the breeding and survival of insect vectors such as mosquitoes. Recent studies suggest that climate change could increase the proliferation of the *aedes* mosquito (vector for dengue), exposing an additional 2 billion people to dengue transmission by the 2080s (Hales et al. 2002). Direct effects from heat waves can cause skin cancer, cataracts, sunstroke and reduced efficiency of the immune system (McMichael et al. 2006).

While climate change tends to be discussed from a global perspective, the health effects are regional, local, and population-specific, and are usually not evenly distributed. There are some communities and population groups that are particularly vulnerable. For example, the United Nations Intergovernmental Panel on Climate Change identified Indigenous groups and coastal communities as two groups that are most vulnerable to climate change. As has been discussed in the chapter on Indigenous health, the close affiliation between most Indigenous people and the natural environment, together with poor socio-economic conditions, predispose them to severe impacts resulting from climate change. Similarly, people living in coastal areas and floodplains are extremely vulnerable to extreme weather events, which can destroy infrastructure and displace entire communities. Temporary relocation of displaced people can lead to increased incidence of communicable diseases due to overcrowding, limited health services, lack of clean water and sanitary facilities, poor mental health and poor nutrition.

1.8 Wars, Conflicts and Health

The unfortunate circumstances of armed conflict and war adversely impacts surrounding ecosystems and affect human health. The settings in which conflicts take place fragment ecosystems, disrupt ecosystem functioning and predispose people to new disease pathogens and new infectious diseases. In addition, the mass fleeing and displacement of people from their communities, force them to live in crowded spaces and under unhygienic conditions, which provide ideal conditions for the onset of infectious and communicable diseases. The limited health care services in refugee camps are usually not adequate to address the myriad health concerns presented, and sometimes these living conditions result in the outbreak of epidemics. Two examples that are noteworthy relate to the emergence of Marburg haemorrhagic fever in Angola, which affected over 200 people, and killed over 90 of the victims (WHO 2007), and the cholera epidemic in the Democratic Republic of the Congo, which killed over 50,000 people.

Marburg haemorrhagic fever, which is related to Ebola, occurred between 2004 and 2005, following a 27-year civil war (1975–2002) in Angola and is reported as the largest epidemic on record (WHO 2007). The disease proliferates in overcrowded areas and settings with inefficient health care services. On the other hand, the cholera epidemic in the Democratic Republic of the Congo, occurred following the Rwandan conflict in 1994. Between 500,000 and 800,000 people fled to seek refuge in the neighboring Congolese city of Goma, when the epidemic struck. The epidemic, which is said to have resulted from a combination of cholera and shigella dysentery, was very fatal, recording a high crude mortality rate of 20–35 per 10, 000 per day (ibid).

1.9 Conclusion

This chapter illustrates the various ways in which human activities impact and, in turn, are impacted by ecosystem dynamics. Ecosystems provide services that are essential for life. These services are continuously under pressure given the growing demand for food and other societal needs. In an attempt to increase productivity, human activities transform ecosystems in ways that compromise their ability to continue to provide ecosystem services. They also transform ecosystems in ways that engender disease and adversely impact both ecosystem and human health. The causal pathways between human activities, their driving forces, and how they transform ecosystems to adversely impact health are complex and not amenable to linear processes. In addition to biophysical processes, social, political, economic and cultural factors further confound these interactions. Hence, given the increasing realization that over the next few decades, the most important determinant of human health will be ecological factors, it is probably prudent for researchers to begin to unravel these intricate connections between society, health, and environment, and ensure that power relations, and social and political considerations are incorporated into people-environment-health analysis. Such understanding will help develop appropriate interventions that will be socio-politically acceptable and also biophysically relevant.

The role of ecological factors as important determinants of human health is not new, but dates back to the 19th century. Interests in ecological factors were superseded by modern medical techniques such as the discovery of microbes, viruses, DNA and the increasing focus on individual lifestyle factors. With issues such as climate change and the rapid emergence of new diseases with mediated by ecological factors, there is, once again, growing interests in the use of ecological approaches to public health, with particular emphasis on ecosystem-human dynamics. For the past few decades there have been growing efforts to integrate health and environment concerns and to develop more ecological and holistic approaches to public health improvement, and sustainable natural resources management. This trend has given rise to new approaches such as the ecosystem approaches to human health, also known as the Ecohealth approach. Before discussing some of the key elements of this approach, it is important to trace the evolution of events leading to a renewed interest in ecological approaches to health, and in particular, the ecosystem approach to human health (Forget and Lebel 2001).

References

Bryant RL (1998) Power, knowledge and political ecology in the Third world. Prog Phys Geogr 22:79–94

Bryant RL, Bailey S (1997) Third world political ecology. Routledge, London

Coluzzi M (1994) Malaria and the Afro-tropical ecosystems:impact of man-made environmental changes. Parassitologia 36:223–227

Colwell RR (1996) Global climate and infectious disease: the cholera paradigm. Science 274:2025–2031

Daszak P, Cunningham AA, Hyatt AD (2001) Anthropogenic environmental change and the emergence of infectious diseases in wildlife. Acta Trop 78:103–116

Delgado C, Rosegrant M, Steinfeld H, Ehui S, Courbois C (1999) Livestock to 2020: the next food revolution. Food, agriculture and the environment discussion Paper 28. 2020 Vision. International Food Policy Research Institute, Washington, DC, 83 pp

FAO/WHO (2002) Global Forum on Food Safety Regulators, Marrakech, Morocco, 28–30 January 2002: Japanese encephalitis/Nipah outbreak in Malaysia. Rome, Food and Agriculture Organization, 2002

Forget G, Lebel J (2001) An ecosystem approach to human health. Int J Occup Environ Health 7 (2 Suppl):S3–S38

Forsyth T (2003) Critical political ecology: The politics of environmental science. Routledge, London

Fostier AH, Forti MC, Guimaraes JR, Melfi AJ, Boulet R, Espirito Santo CM et al (2000) Mercury fluxes in a natural forested Amazonian catchment (Serra do Navio, Amapa State, Brazil). Sci Total Environ 260:201–211

Garofalo C, Vignaroli C, Zandri G et al (2007) Direct detection of antibiotic resistance genes in specimens of chicken and pork meat. Int J Food Microbiol 113(1):75–83

Glass GE, Schwartz BS, Morgan JMIII, Johnson DT, Noy PM, Israel E (1995) Environmental risk factors for Lyme disease identified with geographic information systems. Am J Public Health 85:944–948

Graham JP, Leibler JH, Price LB et al (2008) The animal-human interface and infectious disease in industrial food animal production. Rethinking Biosecur Biocontainment Public Health Rep 123:282–299

Gubler DJ (2002) The global emergence/resurgence of Arboviral diseases as public health problems. Arch Med Res 33:330–342

Hales S et al (2002) Potential effect of population and climate changes on global distribution of dengue fever: an empirical model. Lancet 360:830–834

Harb M, Faris R, Gad AM, Hafez ON, Ramzy R, Buck AA (1993) The resurgence of lymphatic filariasis in the Nile delta. Bull WHO 71:49–54

Kalliola, R, Flores Paitán, S (eds) (1998) Geoecología y Desarrollo Amazónico. Estudio Integrado en la Zona de Iquitos, Perú. Sulkava, Peru:Finnreklama Oy.

Keiser J, Maltese MF, Erlanger TE, Bos R, Tanner M, Singer BH, Utzinger J (2005) Effect of irrigated rice agriculture on Japanese encephalitis, including challenges and opportunities for integrated vector management. Acta Trop 95:40–57

Lam SK, Chua KB (2002) Nipah virus encephalitis outbreak in Malaysia. Clin Infect Dis 34 (Suppl 2):S48–S51

Lebel J, Mergler D, Branches F, Lucotte M, Amorim M, Larribe F et al (1998) Neurotoxic effects of low-level methylmercury contamination in the Amazonian Basin. Environ Res 79: 20–32

Leroy EM, Rouquet P, Formenty P, Souquière S, Kilbourne A, Froment JM, Bermejo M, Smit S, Karesh W, Swanepoel R, Zaki SR, Rollin PE (2004) Multiple Ebola virus transmission events and rapid decline of central African wildlife. Science 303:387–390

Levins R et al (1994) The emergence of new diseases. Am Sci 82:52–60

Mahmood F, Chiles RE, Fang Y, Barker CM, Reisen WK (2004) Role of nestling mourning doves and house finches as amplifying hosts of St. Louis encephalitis virus. J Med Entomol 41: 965–972

McMichael AJ (1997) Global climate change: the potential effects on health. Br Med J 315: 805–809

McMichael AJ, Woodruff RE, Hales S (2006) Climate change and human health: present and future risks. Lancet 367:859–869

Millennium ecosystem assessment (2005) Ecosystems and human well-being: health synthesis. World Health Organization, Geneva

Molyneux DH (1998) Vector-borne parasitic disease – an overview of recent changes. Int J Parasitol 28:927–934

Patz JA, Daszak P, Tabor GM et al (2004) Unhealthy landscapes: policy recommendations on land use change and infectious disease emergence. Environ Health Perspect 112:1092–1098

Patz JA, Graczyk TK, Geller N, Vittor AY (2000) Effects of environmental change on emerging parasitic diseases. Int J Parasitol 30:1395–1405

Peet R, Watts M (1996) Liberation ecology: development, sustainability and environment in an age of market triumphalism. In: Peet R, Watts M (eds) Liberation ecologies: environment, development, and social movements Routledge, London, pp 1–45

Prüss-Üstün A, Corvalan C (2006) Preventing disease through health environments. Towards an estimate of the environmental burden of disease. World Health Organization, Geneva

Schmidt KA, Ostfeld, RS (2001) Biodiversity and the dilution effect in disease ecology. Ecology 82:609–619

Steere AC, Coburn J, Glickstein L (2004) The emergence of Lyme disease. J Clin Invest 113(8):1093–1101

Tadei WP, Thatcher BD, Santos JM, Scarpassa VM, Rodrigues IB, Rafael MS (1998) Ecologic observations on anopheline vectors of malaria in the Brazilian Amazon. Am J Trop Med Hyg 59:325–335

Taylor LH, Latham SM, Woolhouse ME (2001) Risk factors for human disease emergence. Philos Trans R Soc Lond B Biol Sci 356:983–989

Thompson DF, Malone JB, Harb M, Faris R, Huh OK, Buck AA et al (1996) Bancroftian filariasis distribution and diurnal temperature differences in the southern Nile delta. Emerg Infect Dis 2:234–235

UN-Habitat (2006) State of the world's cities 2006/07. The Millennium Development Goals and Urban Sustainability: 30 Years of Shaping the Habitat Agenda. Earthscan, London. 204 pp

Wilson ME (2000) Environmental change and infectious diseases. Ecosyst Health 6:7–12

Wolfe ND, Eitel MN, Gockowski J, Muchaal PK, Nolte C, Prosser AT et al (2000) Deforestation, hunting and the ecology of microbial emergence. Global Change Hum Health 1:10–25

World Health Organization Fact Sheet No. 262. (2009) Nipah virus

World Health Organization (1948) Preamble to the Constitution of the World Health Organization as adopted by the International Health Conference, New York, 19 June – 22 July 1946; signed on 22 July 1946 by the representatives of 61 States and entered into force on 7 April 1948

World Health Organization (1986) Health Promotion: Ottawa Charter. International Conference on Health Promotion, Ottawa, 17–21 November 1986. Geneva, Switzerland

World Health Organization (2007) The world health report – A safer future: global public health security in the 21st century

World Health Organization (2008) The world health report. Primary health care (Now more than ever)

Chapter 2
Evolution Towards an Ecosystem Approach to Public Health

Contents

2.1 Introduction

This chapter discusses the key milestones leading to the emergence of ecosystem approaches to public health thinking, research and practice. The chapter discusses events in three areas that led to public health thinking toward an ecosystem approach. First, the chapter traces evolution of thinking in public health from the "old" public health, through to the "new" public health, and on to "critical" new public health. The "old" public health is characterized by three phases: the sanitary phase (1840s–1870s), the preventive phase (1870s–1930s), and the therapeutic phase (1930s–1970s). All three phases are associated with the biomedical model of health, with each era defined according to dominant forms of medical knowledge (Brown and Duncan 2002). With growing criticism of the individualistic focus of the biomedical model of health and its failure to respond to the complex and structural determinants of poor health, a "new" public health emerged in the mid-1970s. The focus of this "new" public health was to shift the focus from the individual to a multi-causal, socio-ecological approach to health, taking into account the interaction of social, environmental, psychosocial and other factors in producing ill health. With the emergence of the postmodern era and critical theory, the production and circulation of scientific knowledge claims came under scrutiny. Critical scholars

C.Y. Dakubo, *Ecosystems and Human Health*, DOI 10.1007/978-1-4419-0206-1_2,
© Springer Science+Business Media, LLC 2011

began to challenge the objectivity of scientific knowledge claims, instead seeking to illustrate how such claims could be shaped by socio-political, cultural and historical contexts. Critical scholars see the production of scientific knowledge claims to be closely linked with the exercise of power, and interrogates the assumptions and practices of the new public health movement. In particular, critical scholars compare the principles underlying certain health promotion practices to new forms of governance, regulation and social control (Lupton 1995).

Secondly, within the natural resources management sector, there were growing concerns about the discrete, isolated approaches to managing the social, environmental, and economic components of natural resources, with more emphasis placed on the economic component to the detriment of the other two (Hancock 1990). Instead, an integrated approach that gave equal importance to all three components was recommended, giving rise to an ecosystem approach to natural resource management.

Thirdly, increasing calls for sustainable development and the preservation of the environment complemented events in both the public health and natural resource management sectors. Global initiatives such as the Brundtland Commission, the United Nations Commission on Environment and Development, the Johannesburg Summit on Sustainable Development and the 2010 Summit on Climate Change in Copenhagen, all buttress the need for holistic approaches to promoting sustainable human development, which incorporates both social and environmental dimensions. Together, these efforts are intended to respond to health, environment and sustainable development concerns in a concerted manner, and to simultaneously promote human health and ecosystem health.

2.2 "Old" Public Health and the Biomedical Approach

Ecological approaches to public health research and practice date back to the early nineteenth century with the emergence of the sanitary paradigm in Europe. Occurring between the 1840s and 1870s, diseases, especially those in the urban slums of European cities, were attributed to contaminated environments. The external environment was considered a filthy place, filled with dirt, pestilence, and contaminated water, soil and air. This polluted environment was thought to be responsible for major disease epidemics, although there was no scientific evidence to back this claim (Pedersen 1996). The primary focus of public health at the time, then was to monitor the transfer of dangerous substances from the physical environment, including air, water, and food, into the human body, and those being excreted from the human body, including urine, faeces, sputum, and semen into the environment. The control of diseases focused on environmental remediation and involved proper garbage disposal, closed drainage and sewage systems, and the adoption of hygienic practices and behaviour (Dubos 1968). The control of diseases was also broadened to include addressing poverty and broader social problems, after Edwin Chadwick and Engels, two pioneers of the sanitary paradigm, argued that poverty and environmental diseases are intricately linked, and effective interventions must be broadened beyond environmental remedies to include broader societal factors (Susser 1987). The sanitary paradigm paved the way for the introduction of fields like hygiene, public works and sanitary engineering.

In the late nineteenth century, this understanding that environmental and broader societal factors were responsible for the causes of diseases was soon superseded by the emergence of the germ theory and the discovery of microbes. This era dominated medical and public health sciences until the mid-twentieth century (1870s–1930s). The germ theory was characterized by ground-breaking work by pioneers like Robert Koch, who demonstrated in 1882, that a mycobacterium (tuberculosis bacilli) was the causal agent responsible for tuberculosis (1912); John Snow's work on cholera, and Louis Pasteur work that demonstrated that a living organism was the agent in an epidemic afflicting silkworms (Susser and Susser 1996). The germ theory postulated that specific micro-organisms were responsible for the causation of specific diseases. Both the environment and the human body were contaminated by invisible micro-organisms that resulted in infectious and parasitic diseases. The germ theory was dominated by cause-effect linkages between microbes and epidemics, and it was believed that the exposure of individuals to certain microbes in a contaminated environment resulted in specific diseases (Pedersen 1996). The focus on specific agents led to an overly reductionist model of disease causation and a narrow laboratory perspective of identifying and experimentally transmitting disease-causing microorganisms (Evans 1976).

At the time, appropriate interventions for the control of infectious diseases included immunizations, personal hygiene and health education. Anti-microbial agents, DDT and other pharmacologically active substances were used to fight micro-organisms and disease vectors such as mosquitoes. With the discovery of antibiotics in the 1940s, coupled with advances in microbiology and laboratory-based science and diagnosis, many were convinced that the fight against infectious and communicable diseases and other global epidemics was over. This was complemented nicely by a reduction in communicable diseases in most developed countries, with infectious diseases such as typhoid, diphtheria and tuberculosis reducing in numbers (Susser and Susser 1996). This optimism was so high that following World War II, some world leaders and international organizations declared the eradication of malaria from the planet (Garret 1994). The germ theory era made way for molecular medicine, as soon as, viruses and genes were detected.

The victory over infectious diseases was short-lived by the emergence of chronic diseases in the western world. Following World War II, diseases such as peptic ulcer, lung cancer and coronary and heart disease afflicted many middle-aged men in the western world (Susser and Susser 1996). The origin of these diseases were unknown and public health researchers and epidemiologists made use of a variety of techniques to determine possible factors that enhanced risks (Morris 1957). These methods of investigation were depicted by the "black box" paradigm, which mostly related exposure to outcome with minimal analysis of possible intervening factors or pathogenesis (ibid).

Studies in the early 1970s illustrated the association between certain lifestyles such as sedentarism, alcoholism, and smoking with non-communicable diseases. A longitudinally study in the mid 1980s confirmed the links between certain ways of living and behaviour, with morbidity experience and life expectancy (Berkman and Breslow 1983). In addition, the emergence of HIV/AIDS in the 1980s further increased the focus on personal behaviour as an important determinant of health. The dominant ideology then was that disease and ill health were products of individual lifestyle choices, and the individual was blamed for engaging in certain health deteriorating behaviours and practices. Health thus became a personal responsibility, and public health analysis focused on individual factors, accompanied by individual level interventions such as behaviour modifications, exercise, and diet regimes. Both health and disease were abstracted from the biophysical environment and the broader social context, while health education was instituted to exhort individuals to engage in appropriate health behaviours.

The biomedical model characterized the above eras of public health, with emphasis being placed on therapy, treatment of infectious diseases, cause-effect mechanisms, and behaviour and lifestyle modifications. This approach to health care was complemented with capital-intensive health care facilities and services, and high health care costs. By the mid to late 1970s, many countries, especially those in the developing world, to experience difficulty with rising healthcare costs, and the ability to continue to sustain high-technology medical care (Davies and Kelly 1993). In addition, the costly and capital-intensive health care strategies were unable to respond to the health needs of many developing country nationals (Doyal 1979; George 1976). Besides, in those regions, there seemed to be growing appreciation of the potential of grassroots efforts to respond to health concerns through community participation and self-reliance strategies, as demonstrated by success stories in places like China and Cuba as well as those about Tanzania's barefoot doctors (Matomora 1986; Navarro 1972; Sidel and Sidel 1973).

During this period, academics and international health professionals began to express concerns about the failure of the dominant biomedical paradigm to address growing health disparities between regions and population groups, as well as effectively respond to the growing complexities of the health problems facing society. Critics argued that the undue emphasis of the biomedical approach on the individual blames the victim, and fails to take into account the social context in which health decisions and actions occur (Minkler 1994; Neubauer and Pratt 1981). Also, there was increasing understanding that many of the underlying causes of poor health stem from factors such as poverty, unequal world order, globalization and regional marginalization. The vulnerable in society, the poor, rural residents and those at the bottom of the socioeconomic ladder continue to bear a disproportionate portion of the burden of disease (Schulz et al. 2002). The onus on individuals to modify their own health practices needs to be examined in light of unequal power relations that constrain access to health enhancing resources and decision-making processes. Social, economic, and cultural constraints, as well as, limited financial resources, time, education, information, social networks, poor housing and toxic neighbourhoods all undermine peoples effort to live healthy lifestyles.

While criticism of the biomedical model increased, other scholars, such as McKeown argued that the reduction in death rates in the western world over the past two centuries was largely due to improvements in the physical environment, such as: increase in food supplies, changes in social and economic conditions, smaller family sizes, preventive measures, and the control of infectious diseases, and not due to advanced medical care and technology alone (McKeown 1971, 1976). McKeown's argument then was that major improvements in health in the nineteenth century were not due primarily to medical interventions and therapeutic efficacy, as was now being espoused, but also due to improvements in the social and environmental determinants of health. McKeown's ideas and analysis were incorporated in a Canadian federal publication called the Lalonde Report, or *A New Perspective on the Health of Canadians* (Lalonde 1974). Also complementing this was the increasing realization that emerging public health challenges such as HIV/AIDS were becoming too complex to be understood and adequately intervened upon from uni-dimensional perspectives. The social and ecological model that was dismissed in the nineteenth century had to be reconciled with once again. Public health researchers began to shift their focus to a multi-causal paradigm that saw diseases, be they infectious, chronic or degenerative, as being the result of a complex interaction among a number of factors, including social, biophysical and psychosocial factors (Brown and Duncan 2002). This shift in thinking, together with the Lalonde report (1974) and other international public health initiatives, such as the Alma-Atta Declaration and the Ottawa Charter for Health Promotion, paved the way for new thinking in public health, dubbed the *"new" public health*.

2.3   The "New" Public Health and Ecological Approaches

As discussed above, one key report that was instrumental in shaping the thinking around the new public health, both in Canada and internationally, was the Lalonde Report. The report suggested that four key factors were instrumental in shaping human health. These included: human biology, social and physical environments, lifestyle, and the organization of healthcare. *Human biology* refers to an individual's genetic baggage that predisposes them to specific diseases; *environment*, broadly used, refers to all the factors external to the human body that can influence health and are completely or partially beyond the individual's control; *lifestyle* refers to the choices that individuals make that influence their health, such as smoking, exercise, consuming excessive amounts of alcohol, nutrition, among others; and finally *health care organization* refers to the quality of services that are accessible or available to individuals or communities through local institutions and other regulatory structures (Lalonde 1974). The Lalonde Report was one of the first documents to emphasize the important role of environmental factors in shaping human health, and a call to broaden the scope of improving public health beyond the traditional biomedical model. The report also emphasized the role of individual behaviours in shaping health outcomes, and recommended that human health can be improved by focusing on environmental actions and the adoption of health-enhancing lifestyles. However, as it turns out, the latter was more in tune with the existing biomedical model, and emphasis was placed on lifestyle modification to the detriment of environmental actions. Hence individual actions such as behaviour modification, exercise, nutrition, and individual habits, were accorded more importance over community-based and environmental approaches to health improvement (Lupton 1994; Minkler 1994; Neubauer and Pratt 1981).

Emulating the Lalonde Report, the United States Department of Health, Education and Welfare, published the first Surgeon General Report on Health Promotion and Disease Prevention, *Healthy People* (United States Surgeon General Report 1979). The report discussed the role of both individual behaviours and environmental factors in influencing health. The report drew attention to the "careless habits" of society, poor social conditions, and the continuous pollution of the environment. Just like the Lalonde report, the emphasis seemed to be placed on the "careless habits" and the role of the individual, while ignoring the broader environmental factors (Neubauer and Pratt 1981).

In 1986, the Ottawa Charter for Health Promotion (WHO 1986) was released, and the Charter expressed a new view of health and re-iterated the importance of incorporating ecological factors in health promotion strategies. The Charter acknowledged that ecology, caring, and holism were essential issues in developing strategies for health promotion. The Charter emphasized the interrelations among health improvement, stable ecosystems, the sustainable use of natural resources, and the protection of the environment. It encouraged the conservation of natural resources throughout the world as a global responsibility, and emphasized incorporating the protection of both the natural and built environments into health promotion strategies (WHO 1986). The concept of community involvement, and equal participation by men and women in health promotion strategies were also endorsed in the Charter. The Charter proposed five action areas, including building healthy public policies, creating supportive environments, strengthening community action, developing personal skills, and reorienting health services. According to Kickbusch (1989), the Ottawa Charter for health promotion was the first document to delineate an agenda for the *new public health* by locating within the context of new ecological thinking.

The objective of this new public health has been to re-orient public health from an individualistic focus to a more social and ecological approach; integrating social, environmental, cultural, and community aspects of health (Green et al. 1996; McLeroy et al. 1988, 1992; Stokols 1992). This new public health emphasizes prevention, rather than curative interventions. It is also concerned with the reduction of health disparities among various social groups, the production of healthy living and working environments, and the promotion of community participation and individual empowerment (Brown and Duncan 2002). Health from this new perspective is no longer the absence of disease, but seen as a resource for everyday life (Green 1984). This paradigm shift is closely linked to calls for the use of more community-based participatory research approaches that involve participants in all phases of the research process, so as to raise community consciousness, while empowering people to respond to their own health concerns in a proactive manner (Laverack and Labonte 2000; Robertson and Minkler 1994; Schwab and Syme 1997). It also reflects ideas expressed in earlier international health initiatives such as the Alma-Ata Declaration on Primary Health Care (WHO 1978). For example, the Alma-Ata Declaration encourages balancing medical approaches to health care with greater emphasis on the social, political and environmental determinants of health, especially for vulnerable and disadvantaged populations. The Declaration also identifies adequate nutrition, safe water, and basic sanitation as part of a number of essential elements for health improvement. In addition, the Declaration gave formal recognition to the role of community participation in health and encouraged a shift away from health-sector to multi-sector or inter-sectoral approaches to health intervention.

Following the release of the Ottawa Charter, subsequent initiatives expressed similar ecological sentiments to health promotion. Among these are: the "Health for All by the Year 2000" strategy, Healthy Public Policy Conference in Adelaide, Call for Action: Health Promotion in Developing Countries, the Healthy Cities Project Initiatives, and the Sundsvall Statements on Supportive Environments. For example, the Healthy Cities Project Initiative emphasizes the intricate connections between human health and the quality of the environment in which people live (Green et al. 1996). Cities seeking to become healthy are encouraged to engage in intersectoral planning and decision-making processes to identify healthy choices for their communities (Flynn 1996). Throughout Canada, a number of cities have designated themselves as "healthy communities," emphasizing active community participation, intersectoral collaboration, and mutual dependence between the individual and the broader society (Minkler 1999). Also, in 1991, the third international conference on Health Promotion, held in Sundsvall, Sweden, stressed the intricate linkages among health, environment, and human development, and emphasized how development activities must strive to improve both quality of life and health, while preserving the sustainability of the environment (Sundsvall Conference 1991). Recent initiatives such as, the United Nations Millennium Development Goals, 2002 Summit on Sustainable Development in Johannesburg, and the Johannesburg Water, Energy, Health, Agriculture, and Biodiversity (WEHAB) Framework, Millennium Ecosystem Assessment, and the 2010 Climate Change Summit in Copenhagen have all drawn attention the linkages between environment and human health, and the need for broader ecological approaches that span beyond the health sector. At the continental level, there have been on-going regional meetings between environment and health ministers in Africa, Europe, and Latin America. Together these initiatives have paved the way for modeling and thinking about public health from an holistic, ecological and integrated perspective.

2.4 Integrated Approaches to Natural Resource Management

While the health sector sought to develop an inclusive and ecological approach to improving health, there were growing concerns about role of human activities in causing environmental degradation, resource depletion, and climate change. There was also growing disillusionment with the conventional approach to natural resource management that discretely managed resources to satisfy isolated community, economic and environmental objectives, without taking into consideration how such objectives are intertwined and as such should be accorded equal importance. In a schematic developed by Hancock (1990), the conventional approach to natural resource management focuses excessively on managing natural resources to satisfy economic goals, to the detriment of the social and ecological goals (ibid). Such an approach fails to see human beings as integral to the ecosystem. It also fails to recognize that the overall sustainability of an ecosystem resides in balancing the social, economic and environmental aspirations.

One area where the concern for the adoption of an integrated approach to natural resource management first surfaced was in the Great Lakes Basin shared by Canada and the United States. Following World War II, it was fasely assumed that, by the shear vastness of the Great Lakes Basins, it would be able to assimilate toxic substances, and as such became the dumping ground for toxic effluents (Great Lakes Research Advisory Board 1978). However, following a study it was observed that the aquatic ecosystem together with its fauna and flora had become extensively degraded (Colborn 1996). This realization led to the establishment of the International Joint Commission by both the United States and Canadian governments with the mandate to examine how to best manage the watershed in ways that would allow the continuous use of its resources for both social and economic purposes while preserving the integrity and sustainability of the biophysical characteristics of Great Lakes Basin (Great Lakes Research Advisory Board 1978; International Joint Commission 1991). In response, the International Joint Commission (IJC) proposed an "ecosystem approach" to watershed management. The ecosystem approach seeks an integrated approach to natural resource management that takes into account all the key elements of a particular ecosystem, including air, water, land, fauna and flora, and also the human inhabitants. The approach takes into account the intricate linkages between the biophysical ecosystem, economic activity, and human health concerns, and seeks to balance these so as to achieve sustainable development (Rapport 1995). Rather than manage these issues in isolation, the ecosystems approach makes use of a systems perspective, which views human needs, economic, and ecological goals as intricately bound and needs to be addressed as an integrated whole. The ecosystem approach situates human beings at the core of the ecosystem, sees human health as integral to healthy ecosystems, and so ensures ecosystem management contributes positively to the health of both ecosystems and human beings (Forget and Lebel 2001). The Great Lakes scientists were among the first scientific group in North America to propose an ecosystem approach to resource management (International Joint Commission 1991).

Since its application, the ecosystem approach has been endorsed and adapted by a variety of groups, including the Canadian Council of Ministers of the Environment, the Convention on Biological Diversity, and the Millennium Ecosystem Assessment. The Canadian Council of Ministers of the Environment (CCME 1994) describes the ecosystem approach as:

viewing the basic components (air, land, water, and biota – including humans) and functions of ecosystems in a broad context, integrating environment, social, and economic concerns (p. 3).

The Convention on Biological Diversity defines the ecosystem approach as follows:

> The ecosystem approach is a strategy for the integrated management of land, water and living resources that promotes conservation and sustainable use in an equitable way. . . . the application of the ecosystem approach will help to reach a balance of the three objectives of the Convention: conservation; sustainable use; and the fair and equitable sharing of the benefits arising out of the utilization of genetic resources. An ecosystem approach is based on the application of appropriate scientific methodologies focused on levels of biological organization, which encompass the essential structure, processes, functions and interactions among organisms and their environment. It recognizes that humans, with their cultural diversity, are an integral component of many ecosystems.

Given the varying needs and competing interests of stakeholders and the complexity of ecosystem structure and function, the application of an ecosystem approach requires the integration of a variety of perspectives. It is important to integrate expertise from a variety of disciplines including, economics, medicine, anthropology, sociology, veterinary sciences, and ecology (Rapport 1995; Rapport et al. 1999), as well as, lay perspectives and local knowledge from inhabitants and resource users. By integrating across the natural, social, and health sciences, the ecosystem approach transcends disciplinary boundaries and brings together the unique views and knowledges of the various disciplines, and allows for a nuanced understanding of the complexities surrounding ecosystem use and management (Rapport 1995; Rapport et al. 1999).

2.5 Making the Links with Sustainable Development

Alongside the ecosystem approach to natural resource management, there have been increasing efforts to make explicit the links among health, environment, and sustainable development. For example, in 1987 the World Commission on Environment and Development introduced the concept of sustainable development in a report entitled *Our Common Future* (Brundtland 1987). The Commission defined sustainable development as:

> a process of change in which the exploitation of resources, the direction of investments, the orientation of technological development, and institutional change are all in harmony and enhance both the current and future potential to meet human needs and aspirations.

Although the report did not single out human health as one of the key areas of focus, it identified society's role in changing the environment, and how these environmental changes, in turn, affected human health. The Chair of the Commission, former Prime Minister of Norway, Mrs. Gro Harlem Brundtland, later explained that it was not important to single out human health as an area of emphasis, when the entire report was about health (WHO 1998). In essence, the commission saw health to be a central goal of human development, and the protection of the environment and the improvement of human health must be addressed conjointly.

Following on this path, in 1992 the United Nations Commission on Environment and Development (UNCED 1992), held an Earth Summit in Rio de Janeiro which drew further attention to the global deterioration and depletion of the world's ecological resources and the potential implications for human health. The report drew attention to how so-called development programs, underdevelopment, and poor development practices, could deteriorate the environment and negatively impact human health. The human dimension of sustainable development was emphasized through the first principle of the Rio Declaration, which stated that "....human beings are at the centre of concerns for sustainable development. They are entitled to a healthy and productive life in harmony with nature." Agenda 21, the action plan for UNCED, took this further by stressing that the health needs of the world's population need to be addressed urgently through strategies that would acknowledge the interconnections among all dimensions of the environment, development and human health. For the first time, the need for a concerted, transdisciplinary approach to improving human health in the context of environmental sustainability was recommended. Health was no longer an issue for only the medical community, but also for professionals within agriculture, housing, public works, sanitation, and natural resource management.

Following the Earth Summit, the World Health Assembly in 1992 formulated a new WHO Global Strategy for Environmental Health, partially taking into account the recommendations and new thinking around health from the Earth Summit (WHO 1998). The Strategy articulated this new thinking as follows:

(1) Health is a an essential component of sustainable development which can only be achieved through concerted action by all sectors of society;
(2) action in the physical and social environments to improve health is taken in close partnership between the health sector and those other sectors, which have a strong impact on environmental quality;
(3) health is also affected by the actions of individuals, families, community groups that have an enormous impact on their environments

Recent reports such as the Health Synthesis of the Millennium Ecosystem Assessment (2005) also draw attention to the links between ecosystems and human health.

2.6 Modeling Human Health from an Ecosystem Perspective

Recent events in both the public health and natural resources sectors, together with global initiatives on sustainable development have rekindled interests in modeling human health from an ecosystem perspective (Van Leeuwen et al. 1999). Modeling human health from an ecosystem perspective places human beings squarely at the centre of ecosystem management, suggesting that human beings, either individually or collectively, influence and are influenced by the surrounding biophysical, social, and economic environments, and the existing policies governing such environments (ibid). Van Leeuwen and colleagues (1999) reviewed a number of ecosystem models of human health from the late nineteenth century to the 1990s, and eventually proposed a "butterfly model of health", which took into account key features and attributes of past models. The butterfly model reflects the complex ways in which key elements of the biophysical and socio-economic environments of humans interact within an ecosystem context. The authors demonstrate how the model can be applied to human populations assembled according to political boundaries (e.g. communities, provinces) or ecological boundaries (watersheds, farmlands).

The concept of an ecosystem serves as a useful construct for illustrating the complexities and interactions of the myriad factors influencing health from varied temporal and spatial dimensions. Ecosystems exist in multiple spatial and temporal dimensions, and are usually thought of as being organized in nested hierarchies, with each level of the hierarchy demonstrating inherent properties that occur as a result of the complex interactions of the many internal and external components and functions of the ecosystem (Van Leeuwen et al. 1999).

Ecological approaches to health tend to seek a balance among individual level factors and broader social and ecological factors. Hence modeling human health from an ecosystem perspective makes use of systems thinking that recognize human health as influenced and conditioned by factors at various levels and scales including those related to the individual, family, community, surrounding biophysical and socio-economic environments, and national and global policies. This nested nature of health determinants calls for an examination of how factors at these various levels interact and influence health outcomes, and helps identify appropriate levels and targets for intervention. This ecosystem approach to health recognizes that the complexity and multiplicity of factors influencing human health cannot be adequately resolved thorough uni-dimensional or piece meal approach. Instead, health must be promoted from an holistic and integrated perspective. Ecological approaches try to illustrate the reciprocal nature of the relationships between people and their environments, and place emphasis on the fact that improved health is achieved through concerted efforts between the intrapersonal level and broader community, institutional, and policy factors (Bronfenbrenner 1990).

One of the earliest ecological models of health to depict this thinking, is the "Mandala of Health" (Hancock and Perkins 1985). A *mandala* represents a circular design of concentric rings incorporating multiple factors ranging from the biological and personal to the biosphere. Individual health is situated in the centre, and is comprised of the mind, body and spirit. This initial ring is then influenced by circular nested systems of the household, the community, the human-made environment, culture, and the biosphere. These nested rings are considered to be intimately interdependent and jointly influenced by other social and political forces. They are also dynamic in size and shape, depending on the temporal and spatial contexts (Van Leeuwen et al. 1999). In addition, there are a number of factors that have to be taken into account when analyzing how individual health is shaped by all these nested rings including: individual developmental histories and social support systems; community-mediating structures, such as community networks and power structures; access to, and control of community and ecosystem resources; organizational structures and processes that can negatively influence health; participation in decision making processes; advocacy, and content of public policies (Minkler and Wallerstein 2003).

There have also been health models from the perspective of a community ecosystem. For example, Hancock (1990) proposed a community ecosystem model of health which integrates the concepts of health and sustainable development in the context of the community. This model complements the Mandala of Health model and is suitable for communities striving to become both healthy and sustainable. The community ecosystem model is comprised of three overlapping circles: community aspirations, the economy, and the environment. Centrally located in the middle of the three overlapping circles is "health or human development." Hancock suggests that for human health development to be optimized, three qualities must be met in each for the three circles: community, economy, and environment. For example, within the community, it is important that the community be convivial; that is have social support networks, provide opportunities for community members to live together in harmony and be able to participate fully in decision-making processes. Also, the built environment of the community must be liveable; that is possesses an urban structure that supports conviviality and also provides a viable human environment. Lastly, the community must be equitable, ensuring that its members are treated fairly, that people are able to meet their basic necessities, and have equal opportunities to reach their optimal potential.

With respect to the economy, the main requirement is that the economy is adequate and able to generate sufficient wealth to enable community members obtain a satisfactory level of health. Economic wealth must be equitably distributed within the community, and the economy must be environmentally sustainable.

For the environment, the primary requirement is that the environment be sustainable over the long term, be viable for humans, and be able to provide clean air, water, and food. Also the environment must be perceived as comprising of both the built and natural environments, and must be liveable from a community and human perspective (Hancock 1993). In addition, it is important to note that in order to achieve sustainable health development, issues in the three circles of environment, community, and economy must be addressed in an integrated manner, not in isolation or piece mealy.

Modeling human health from an ecosystem perspective is designed to overcome the shortcomings of past socio-ecological approaches to health. These approaches were criticized for the lack of centrality of ecological factors. For example, although reference is usually made to the "environment" as an important determinant of health, most emphasis has always been placed on the social determinants. A renewed focus on ecosystem approaches to public health concerns is timely, given the increasing attention of the role of ecological factors on human health, the rapid pace of emerging new diseases, and the growing concerns about climate change. The uniqueness of the ecosystem approach to public health is that, it is not only interested in improving human health, but emphasizes achieving this through the sustainable management of the environment.

The ecosystem approach has been adopted by a number of institutions world-wide and is being taught at a number of universities, and also incorporated into a number of medical school curricula. For example, in Canada some of these univer-sities include the University of Western Ontario, University of British Columbia, and University of Guelph. Also in Canada, the International Development Research Centre (IDRC) is a pioneer in the application of ecosystems approaches to human health. IDRC makes use of the ecosystem approach to promote health in many developing countries. Given that many of the health problems in developing coun-tries have ecologically components, it probably makes sense from a cost-effective to encourage a wider adoption of the ecosystem approach in many public health settings, as well as integrate it into national health and environment policies. The First Inter-Ministerial Conference of Health and Environment Ministers in Africa, held in Gabon in 2008, saw many African ministers endorse the ecosystem approach to human health as a useful approach to help curtail most of the environmentally-mediated health problems in the region. The core concepts and principles of the ecosystem approach to health will be discussed in the next chapter.

2.7 Towards Critical Public Health

Prior to discussing the ecosystem approach to human health, it is important to draw attention to the continuous evolution of public health thinking to incorporate crit-ical social theoretical perspectives that are emerging from other disciplines such as human geography anthropology, sociology, and education. Critical public health seeks to examine the underlying assumptions, practices, and knowledge claims of the new public health, movement, by placing these claims in an historical and socio-political context (Lupton 1998). This body of knowledge is influenced by critical social theory, discourse theory, and the sociology of science. Scholars influenced by critical perspectives (referred to in this book as critical scholars) view knowl-edge as socially constructed and mediated through perspectives of the dominant society. They argue that knowledge is always partial and situated within particular systems of meanings and epistemological positions (Nicholson 1990). This is partic-ularly evident within medicine, and especially public health and health promotion. A critical perspective in public health queries the taken-for-granted assumptions underlying health knowledge and practices; examines who controls these assump-tions, how public health problems are constructed, defined and explained, and also examines the processes through which alternate views are marginalized and val-orized (Lupton 1998). Critical perspectives call for the use of rigorous analytical frameworks to examining phenomena. For example, within public health, the inves-tigation of the causal factors responsible for ill health must be examined with the contexts of their social, political, and historical framings, to ensure the explication of any hidden agendas that may influence proposed interventions.

Drawing on Foucault's writing on medicine and governmentality, critical public health scholars suggest that western scientific claims, including medical and health knowledge systems have become a primary means of organizing and normalizing peoples behaviors and lives (Turner 1994). This critically informed literature views the discursive practices of the new public health as representing new forms of regulation, governance, and social control. According to Lupton (1995), the discursive practices of the new public health, especially those related to health promotion, legitimize ideologies and social practices through the identification of exercise, diet, and behavioral regimes. Also linked to these discussions are notions of knowledge and power, and how they mediate each other.

Ecohealth as an emerging field is yet to fully benefit from these theoretical developments, and this book makes the first attempt to apply such critical perspectives to the field of ecohealth. Critical public health and its application to the ecohealth approach are discussed in subsequent chapters.

2.8 Conclusion

This chapter has traced the evolution of public health thinking from the sanitary paradigm through to the postmodern perspectives of critical public health. As illustrated above, ecological thinking in public health is not entirely new, it is re-emerging as scholars take interest in issues such as climate change, environmental degradation and how environmental conditions mediate newly emerging diseases. This renewed interest has also spawned new approaches such as the ecosystem approach to human health, which is gaining widespread attention among institutions, academia, public health practitioners. With recent developments in the application of critical theory to public health, it is anticipated that ecohealth will benefit from these developments and emerge as a theoretically rigorous field of study.

References

Berkman LF, Breslow L (1983) Health and ways of living: the Alameda county study. Oxford University Press, New York

Bronfenbrenner U (1990) The ecology of human development: experiments by nature and design. Havard University Press, Cambridge, MA

Brown T, Duncan C (2002) Placing geographies of public health. Area 33:361–369

Brundtland G (1987) Our common future. Oxford University Press, Oxford

Canadian Council of Ministers of the Environment (CCME) (1994) A framework for developing goals, objectives and indicators of ecosystem health: tools for ecosystem management. CCME, Ottawa, ON

Colborn T (1996) The great lakes: a model for global concern. In: Di Giulio RT, Monosson E (eds) Interconnections between human and ecosystem health. Chapman & Hall Ecotoxicology Series, London, pp 85–91

Davies JK, Kelly MP (1993) Introduction. In: Kelly MP, Davies JK (eds) Healthy cities: research and practice. Routledge, London, pp 1–13

Doyal L (1979) The political economy of health. Pluto, London

Dubos R (1959) The mirage of health: Utopias: progress and biological change. Harper & Row, New York

Dubos R (1968) Man, Medicine and environment. Mentor books, New American Library, New York

Evans AS (1976) Causation and disease: the Heule Koch Postulatis revisted. Yale J Biol Med 49:175–195

Flynn BC (1996) Healthy cities: toward worldwide health promotion. Annu Rev Public Health 17:299–309

Forget G, Lebel J (2001) An ecosystem approach to human health. Int J Occup Environ Health 7(2 Suppl):S3–S38

Garret L (1994) The coming plaque: newly emerging diseases in a world out of balance. Farrar, Strauss and Giroux, New York

George S (1976) How the other half dies: the real reasons for world hunger. Penguin Books, Harmondworth, Middlesex

Great Lakes Research Advisory Board (1978) The ecosystem approach: scope and implications of the an ecosystem approach to transboundary problems in the Great Lakes basin. Special Report to the International Joint Commission

Green LW (1984) Modifying and developing health behaviour. Annu Rev Public Health 5:215–236

Green LW, O'Neill M, Westphal M, Morisky D (1996) The challenges of participatory action research for health promotion. Health Promotion Educ 3:3–5

Hancock T (1990) Healthy and sustainable communities: health, environment and economy at the local level. A presentation at the 3rd Quebec Colloquium on Environ- ment and Health, Quebec City, November. York

Hancock T (1993) Health, human development and the community ecosystem: three ecological models. Health Promot Int 8:41–47

Hancock T, Perkins R (1985) The Mandala of health: a conceptual model and teaching tool. Health Educ 24:8–10

International Joint Commission (1991) A proposed framework for developing indicators of ecosystem health for the Great Lakes Region. IJC, Windsor, ON

Kickbusch I (1989) Good planets are hard to find – Approaches to an ecological base for public health. In: Brown V (ed) A sustainable healthy future: toward an ecology of health. La Trobi University and Commission for the Future, Melbourne, pp 7–30

Lalonde M (1974) A new perspective on the health of Canadians. Ministry of Supply and Services, Canadian Federal Government, Ottawa, ON

Laverack G, Labonte R (2000) A planning framework for community empowerment goals within health promotion. Health Policy Plan 15(3):255–262

Lupton D (1994) Medicine as culture. Sage, London

Lupton D (1995) The imperative of health. Public health and the regulated body. Sage, London

Lupton D (1998) The emotional self: A socio-cultural exploration. Sage, London

Matomora MK (1986) A people-centered approach to primary health care implementation in Mvumi, Tanzania. Social Sci Med 28:1031–1037

McKeown T (1971) An historical appraisal of the medical task. in medical history and medical care. Oxford University Press, Oxford

McKeown T (1976) The role of medicine: Dream, mirage, nemesis? Nuffield Provincial Hospital Trust, London

McLeroy KR, Bibeau D, Steckler A, Glanz K (1988) An ecological perspective on health promotion programs. Health Educ Quart 15:351–378

McLeroy KR, Steckler AB, Goodman RM, Burdine JN (1992) Health education research, theory, and practice: future directions. Health Educ Res Theory Pract 7:1–8

Millennium Ecosystem Assessment Series (2005). Ecosystems and human well-being: a framework for assessment; ecosystems and human well-being: Available online at: http://www.millenniumassessment.org/, accessed 10 May 2010

Minkler M (1994) Challenges for health promotion in the 1990s: social inequities, empowerment, negative consequences, and the common good. Am J Health Promot 8(6):403–413

Minkler M (1999) Personal responsibility for health? A review of the arguments and the evidence at century's end. Health Educ Behav 26:121–141

Minkler M, Wallerstein N (eds) (2003) Community based participatory research for health. Jossey-Bass, SanFrancisco, CA.

Morris JN (1957) Uses of epidemiology. Churchill Livingstone, London

Navarro V (1972) Health, health services, and health planning in Cuba. Int J Health Services 2: 397–432

Neubauer D, Pratt R (1981) The second public health revolution: a critical appraisal. J Health Polit Pol Law 6(2):205–228

Nicholson L (ed) (1990) Feminism/postmodernism. Routledge, New York

Pedersen D (1996) Disease ecology at crossroads: man-made environments, human rights and perpetual development utopias. Social Sci Med 43(5):745–758

Rapport DJ (1995). Ecosystem Health: An emerging, integrating science. In, Rapport, D.J., Gaudet C. L., Callow, P (eds) Evaluating and montinoring the health of large-scale ecosystems. NATO ASI Series 1, (28). Springer, Berlin, pp 5–31

Rapport DJ, Bohm G, Buckinham D et al (1999) Ecosystem health: the concept, the ISEH, and the important tasks ahead. Ecosyst Health 5:82–90

Robertson R, Minkler M (1994) New health Promotion Movement: A critical examination. Health Educ Quart 21(3):295–312

Schulz AJ, Krieger J, Galea. S (2002) Addressing social determinants of health: community-based participatory approaches to research and practice. Health Educ Behav 29(3):287–295

Schwab M, Syme SL (1997) On paradigms, community participation and the future of public health. Am J Public Health 87(12):2049–2054

Sidel V, Sidel R (1973) Serve the People. Observations on Medicine in the People's Republic of China. Josiah Macy Jr. Foundation, New York

Stokols D (1992) Establishing and maintaining healthy environments: toward a social ecology of health promotion. Am Psychol 47:6–22

Susser M (1987) Falsification, verification and causal inference in epidemiology: reconsiderations in the light of Sir Karl Popper's philosophy. In: Susser M (ed) Epidemiology, health and society: selected papers. Oxford University Press, New York, pp 82–93

Susser M, Susser E (1996) Choosing a future for epidemiology: eras and paradigms. Am J Public Health 86:668–673

Turner BS (1994) Theoretical developments in the sociology of the body. Aust Cult Hist 13:13–30

United Nations Conference on Environment and Development (UNCED) 1992. The global partnership for environment and development. A guide to agenda 21, Geneva, Switzerland, April 1992

United States Department of Health, Education, and Welfare (1979). Healthy People. The Surgeon General's report on health promotion and disease prevention. DHEW publication No. (PHS) 79-55071. Washington, DC, USA

VanLeeuwen J, Waltner-Toews D, Abernathy T, Smith B (1999) Evolvingmodels of human health toward an ecosystem context. Ecosyst Health 5:204–219

World Health Organization, (1978). Alma-Ata 1978. Primary health care. Geneva

World Health Organization (1991). The third international conference on health promotion: supportive environments for health – the Sundsvall conference. 9–15 June 1991

World Health Organization (1992). WHO World Assembly. Resolution 45.32

World Health Organization (1998) Environmental Health at the Dawn of the Twenty-First: Opportunities and Challenges. Environmental Health News Letter No. 28. Special 50th Anniversary Issue

World Health Organization (1986) Health promotion: Ottawa charter. In: International conference on health promotion, Ottawa, ON, Geneva, Switzerland, 17–21 November 1986

Chapter 3
Ecosystem Approaches to Human Health: Key Concepts and Principles

Contents

3.1 Introduction

The previous chapter discussed some of the key milestones leading to the emergence of the ecosystem approach to human health. This evolution occurred in both the public health and natural resources management sectors, and was buttressed by global initiatives seeking to promote sustainable development. Within the public health sector, the conceptualization of health and its determinants evolved from a narrow, individualistic and biomedical perspective to a broader, ecological and holistic perspective. Similar events in the natural resources sector saw a move towards a more integrated approach to natural resources management, with concerns for human health taking centre stage. Globally, there have been a number of initiatives drawing attention to the interdependencies among society, environment and the health and well-being of individuals and communities. All of these events have dictated a paradigm shift from a sector-based, uni-dimensional approach to human health

C.Y. Dakubo, *Ecosystems and Human Health*, DOI 10.1007/978-1-4419-0206-1_3,
© Springer Science+Business Media, LLC 2011

development to intersectoral, transdisciplinary and integrated approaches to health improvement (See Fig. 3.1).

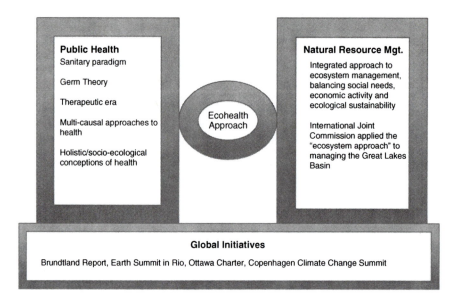

Fig. 3.1 Emergence of ecosystem approaches to human health

With increasing concerns about climate change and ecosystem degradation, and the associated human health implications, it is important that public health interventions refocus their efforts to ensuring that the biophysical environment features prominently as a major determinant of human health. Also, the recent emergence of new diseases such as H1N1 and SARS, illustrate that not only does it take a while to understand the factors responsible for the emergence of these new diseases, but also shows that the actions needed to understand, intervene, and contain their quick spread defy conventional public health approaches. Effective public health strategies require collaborative and integrated efforts from a variety of disciplines, together with cooperation from the public.

The ecosystems approach to human health, or the ecohealth approach is an emerging health promotion strategy designed to shift public health thinking from the traditional, uni-dimensional biomedical approach towards this new transdisciplinary and integrated approach. It makes use of the conceptual construct of an ecosystem to examine the complex and myriad factors influencing human health concerns, and seeks to promote human health and well-being through sustainable management of all components of the environment. The ecohealth approach is premised on the notion that human beings cannot be healthy in an unhealthy environment (Lebel 2003). The health of individuals and their communities are inextricably linked to the health of their biophysical, social and economic environments, and all must be examined in tandem.

The ecohealth approach provides a framework from which to examine these interrelationships, that is the interrelationships between people and the various dimensions of the environment (social, cultural, physical). The approach also allows for the evaluation of the various mechanisms through which both ecosystems and human health are impacted through ecosystem transformations, and then allow for the development of appropriate interventions aimed at improving human health through sustainable ecosystem management. The ecohealth approach can be seen as a strategy that pursues the promotion of human health within the broader context of ecosystem health. Building on a variety of definitions (Forget and Lebel 2001; Rapport et al. 1998), an ecosystem approach to human health can be defined as:

> an integrated approach to investigating the critical links among human activity (social, political, physical), ecosystem conditions (natural or anthropogenic) and human health outcomes, and using this understanding to develop interventions aimed at improving human health through sustainable ecosystem management.
>
> The approach makes use of collaborative processes to understand human-environment dynamics, and integrates knowledge from various disciplinary backgrounds, including the health, natural and social sciences, as well as traditional knowledge systems of local actors so as to allow for a comprehensive understanding of the various components of the ecosystem, the human-ecosystem interactions, and how these interactions shape health outcomes among different societal groups.

The ecohealth approach considers human health and ecosystem health as components of a complex system, and views people as active players in this space, not passive victims. In contrast to the single disciplinary focus of traditional environmental health approaches, the ecohealth approach encourages the investigation of environment and health problems from an intersectoral, transdisciplinary and multistakeholder perspective to allow for a better understanding of the determinants of health and to develop appropriate interventions.

A number of elements are central to the ecohealth approach. The ecohealth approach:

- places equal importance on both human health and ecosystem health, and emphasizes the inherent connections between the two;
- emphasizes the inherent linkages between macro-level and micro-level phenomena, and how these interact at various spatial and temporal scales to influence human health outcomes;
- integrates indigenous/local knowledge systems with knowledge from different disciplines, and other key stakeholders;
- emphasizes meaningful participation by community members and all stakeholders; and
- proposes interventions that seek to simultaneously improve the health of both human beings and the surrounding ecosystems, of which they are integral to.

Prior to elaborating on these issues later on in the chapter, it is important to examine some of the key concepts of the ecohealth approach.

3.2 The Concept of Human Health

The World Health Organization (WHO) defines health as "a state of complete physical, mental and social well-being and not merely the absence of disease or infirmity" (WHO 1948). This definition has come under criticism as being idealistic in its attempt to achieve *a state of complete* health (Noack 1987). Critics argue that this idealistic notion of health can only be approached but never achieved. Others argue that health is not a state but a task; health is a means to an end, and not an end in itself (Duhl 1976; Seedhouse 1986). Despite these criticisms, this definition is still useful for conceptualizing health from an ecosystem perspective. The significance of defining health from this perspective draws attention to health as an holistic, multidimensional phenomenon, that is determined by a multiplicity of factors and focuses on positive well-being, not just the absence of disease or infirmity. The richness of health is taken away when we define health biomedically and focus on conditions of disease, morbidity and mortality. This broader definition of health illustrates health as a social phenomenon, which embodies the quality of our relationships with one another (Labonté 1991).

In 1978, the Alma Ata declaration on primary health care emphasized the social dimensions of health and the importance of community participation in health promotion initiatives (WHO 1978). The declaration encouraged the involvement of people in evaluating their own health concerns so as to blur the boundaries between "experts" and "lay" perspective of health and to increasingly take ownership for their health promotion. Such collaborative evaluation allows health professionals to work with people in the context of their everyday environments (e.g. household, community, workplaces) and to assess how health is shaped in these settings.

In 1986, the Ottawa Charter extended the definition of health to include:

> ..[T]he extent to which an individual or group is able, on the one hand, to realise aspirations and satisfy needs; and, on the other hand, to change or cope with the environment. Health is, therefore, seen as a resource for everyday life, not the objective of living; it is a positive concept emphasizing social and personal resources, as well as physical capacities (WHO 1986: 73).

The Charter viewed health "as a resource for everyday life, not the objective of living." Health is seen as instrumental, a means rather than an end. Accordingly, health is what we need to first possess, in order to accomplish other tasks and aspirations in life. Health is seen as "a positive concept emphasizing social and personal resources, as well as physical capacities" (ibid). The Ottawa Charter for Health Promotion describes the pre-requisites for achieving health to include peace, shelter, adequate education, basic nutrition, sufficient income, a stable environment, sustainable resources, social justice and equity. Achieving health therefore becomes a shared responsibility between society and the individual. Such views of health reflect a change from viewing the individual as a passive victim of health to one playing an active role in shaping their own health outcomes (Lincoln 1992).

The above conceptualizations of health incorporate issues related to ecology, natural resource management, active participation and so seem to align with the holistic philosophy of the ecohealth approach. For example, the Ottawa Charter identifies stable ecosystems and sustainable use of natural resources as important components of health, and presents health as the product of people's continuous interaction with, and interdependence on their environments.

This broad, social and ecological conceptualization of health draws attention to the complexity of factors shaping human health, and the use of an ecosystem as an analytical construct helps illustrate the dynamic interplay of the various factors, and all the feedback mechanisms that shape human health and well-being. As discussed in the previous chapter, an ecosystem approach to health draws attention to how micro-level factors interact with broader environmental factors to shape human-environment relations, and consequently impact human health. It views health from a systems perspective, with human health being influenced by a nested hierarchy of systems at various scales – the family, the community, the culture, the societal structure, and the physical environment. A good understanding of the factors shaping health requires equal attention to both micro-level and macro-level factors, as well, as the dynamic interactions between them, so as to allow for the development of targeted interventions. This conceptualization of health differs from the linear, individualistic ethic that ascribes poor health status to individual characteristics, such as lifestyles choices, behaviours, and habits, without taking into account structural forces that impact health or prevent the achievement of optimal health.

While the World Health Organization has provided a number of formal definitions of health, it is important to realize that health is a social construct and has no pre-determined meanings. The meaning of health and how people experience health are shaped by a variety of contexts, which are constantly changing with changing situations. From a postmodern perspective then, health has no stable or fixed meanings, it is always in flux, and has to defined or interpreted based on context (Fox 1991, 1994). Also, health is perceived, experienced and interpreted differently by different cultures, population groups, and by men and women. As a social construct, then, the definition of health, how it is experienced and intervened upon are subjective and open to debate.

3.3 The Concept of Ecosystem

Within the health and environment literature, the terms "ecosystem" and "environment" always tend to be used interchangeably, as has been the case in this book. However, there are distinctions between the two terms. The term "environment" usually refers to the external or inherent physical conditions that affect and influence the growth and development of organisms. The use of the term in this way portrays environment as existing beyond humans rather than encompassing them (Labonté 1991). The use of "environment" in this manner is consistent with the reductionist approach of investigating environmental health concerns that focus on examining the cause-effect relationships between "proximal" environmental exposures and their associated health impacts. Cultural, political, and social factors are not central in such investigations and are usually treated as confounding variables, rather then essential to shaping human-environment interactions (ibid).

In contrast, the concept "ecosystem" refers to the interaction among a set of living organisms, including humans, and their nonliving environments (Millennium Ecosystem Assessment 2005). Human beings are viewed as integral to ecosystems, and not apart from it, and an ecosystem approach to natural resources management takes into account the social and economic well-being of people. Societal dynamics, including cultural, political, and economic factors play an important role in ecosystem change and influence access to ecosystem resources. These society-ecosystem dynamics directly and indirectly shape health, and so constitute an important component in investigating environmental health concerns.

Tansley first introduced the concept "ecosystem" in 1935 (Tansley 1935). Tansley described the ecosystem as "not only the organism-complex, but also the whole complex of physical factors forming what we call the environment" (Tansley 1935: 299). Tansley noted that ecosystems varied in size and structure. This earlier formulation has been adapted to emphasize the interactions among the living and non-living components of a system (Odum 1953). Hence, ecosystems are usually characterized by a high level of interdependence and interaction between living things and the nonliving components within a defined space in the environment. The "system" in ecosystem simply refers to a set of elements which interact with each other within a defined boundary. Usually the boundary of an ecosystem is defined according to the project or study at hand and could be as small as a farm or as big as an entire continent.

For example, the Canadian Council of Ministers of the Environment (Hancock 1990) note that:

> the limits of a given ecosystem are defined by the user, according to the task at hand and the scope of the process. While in general the limits selected will circumscribe an ecological space such as a watershed or a region, we can also designate a farm, an urban subdivision or a rural community as an ecosystem.

Ecosystems have complex structures and the interactions among the various components of the ecosystem are difficult to characterize or predict given their tendency to vary at different spatial and temporal scales. This complex structure of the ecosystem is sometimes described from the perspective of nested, interlocking hierarchies of geographic units embedded within the biosphere (Kay et al. 1999). Each geographic unit in this nested hierarchy is both a complete entity in itself and part of a larger entity. Each level of the ecosystem hierarchy displays unique characteristics confined to that level, which can modulate each other and at the same time can also influence elements at other levels. Elements at the various levels interlock and influence each other in ways that can negatively or positively influence human health through feedback loops (Waltner-Toews 2001).

This notion of a nested hierarchy is very relevant to the ecohealth approach as it draws attention to the complex mechanisms and processes through which human activities influence and are influenced by factors at various spatial and temporal scales. This concept of a nested hierarchy is useful as an analytical framework to illustrate how individual health is influenced by phenomena at varying scales, ranging from family to the biosphere, with each scale having its own social, economic, cultural, and environmental factors that mediate phenomena at their level and also at other levels (Mergler 2000 unpublished data, cited in Forget and Lebel 2001).

3.4   Assessing the Health of Ecosystems

It is well recognized that the health of human beings depends on the health of the ecosystem that supports them, but just how are we to assess the health of an ecosystem? Like human health, ecologists are concerned with evaluating the state of health of various ecosystems and developing the necessary interventions in a timely and effective manner to prevent the degradation of ecosystems. The health of ecosystems has been described from a number of perspectives, with various definitions incorporating social, economic, human and biophysical dimensions (Costanza 1992; Karr 1991; Kay 1993; Rapport 1992). For example, Constanza and his colleagues (1992) define a healthy ecosystem as one that is able to preserve both its structure and function in light of external pressures over a period of time. Such healthy ecosystem exhibits a number of features, including the ability to be free from "distress syndrome," remain active and maintain its organization, demonstrate autonomy over time, and is resilient to stress. Similarly, Rapport et al. (2001) describe a healthy ecosystem in terms of its: (1) *organization* – the diversity and number of interactions between system components; (2) *resilience* – the capacity of a system to maintain its structure and function in the presence of stress; and (3) *vigor* – the activity, metabolism or primary productivity of a system. Bell (1994) shares a different perspective when pointing out that, the integrity or health of the ecosystem in the twenty-first century could be interpreted as meaning the capacity of nature to continue to serve human beings.

Ecosystem health has also been defined from varying functional perspectives, including the ability to realize inherent potential, capacity to self-repair when damaged, and minimal external resources requirement to maintain sustainability (Karr et al. 1986). The term "integrity" has been used to refer to ecosystem capacity for self-organization and renewal, as well as, the ability of social and economic structures to maintain their organization (Nielson 2001). Nielson (2001) observes that just like human health, it might prove difficult for ecosystems to achieve or maintain this ideal state of health, although it serves as a useful benchmark against which to assess natural and human ecosystem disturbance. From this perspective then, ecosystem health should not be narrowly limited to the absence of disease (distress), but broadened to encompass various aspects of health or sustainability.

While it might be unrealistic to expect ecosystems to remain in pristine states, some ecologists suggest the importance of watching for early signs of distress or "early warning" indicators of sickness. For example, similar to human health, the condition of ecosystems could be monitored through assessment of their "vital signs" including: decline in the size of dominant species (Kerr and Dickie 1984), reversal of trends in ecosystem development (Odum 1985), abnormal ecosystem structures (Schindler 1990), and loss of capacity to recover after an external stress factor (Steedman and Reiger 1990). However, some argue that, by the time these symptoms show up, ecosystem degradation might have begun to take place. Rather than wait for the signs of distress to show up, it is recommended that "early-warning" mechanisms be put in place to monitor key features such as changes in the composition of key biota, and biochemical changes in organisms that, are sensitive to external pressures so as to prevent extensive damage (Maini 1992; Rapport 1992, 1995).

3.5 The Ecosystem Approach to Human Health: Key Issues

As discussed above, there are a number of key issues that are central to the ecohealth approach. One key issue is that, the ecohealth approach views human beings as integral to ecosystems. The approach is based on the premise that, in nature the health and well-being of people cannot be separated from the health of the biophysical environment that sustains life. The interdependencies of human health and ecosystem health is best illustrated through the analogy of the egg white and the egg yoke, whereby an egg can only be good when both the yoke and white are good, so can society be healthy only when both the ecosystem and the people it supports are healthy (International Development Research Centre 1997). Healthy communities are built on the foundation of both healthy populations and healthy natural ecosystems, and these issues must be nurtured conjointly. For example, ecosystem management decisions must take into account aspects of human health and well-being, while health promotion takes into account the social, cultural, and physical aspects of the environment.

Second, the ecohealth approach views human health and ecosystem health as being components of a complex system, and so makes use of systems thinking to explore the relationship between the various components of an ecosystem (human, economic, and environmental), the interactions among them, and how these affect human health. It then evaluates these interactions to identify which determinants of human and ecosystem health to intervene upon (Forget and Lebel 2001). In addition, a systems approach allows for a better understanding of the spatial and temporal dimensions of the factors influencing human health, as well as the inter-linked influences between micro-level and macro-level phenomena impacting health. For example, the ecohealth approach recognizes human health as influenced and conditioned by a nested hierarchy of factors ranging from the individual level to the planetary level, and examines these in light of social and political factors. Individual health is situated in the centre of these nested circles, and is influenced by the individual's unique genetic baggage and lifestyle choices. Broader nested circles of the household, neighbourhood, community, and so on, in turn, influence this individual's health. Also, influencing each circle are social, political, economic forces, including national and global policies (See Fig. 3.2). These nested rings are dynamic

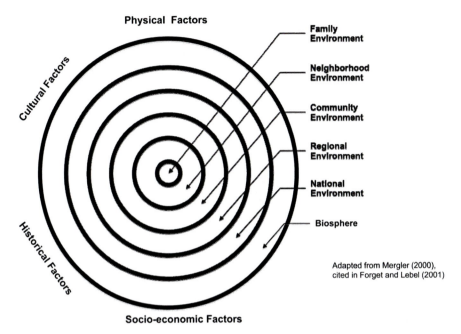

Fig. 3.2   An ecosystem approach to human health

in size and shape, and have temporal and spatial dimensions (Van Leeuwen et al. 1999). A systems approach allows for a better understanding of the interactions of the various determinants of both human health and ecosystem health at various scales, and provides insights for the development of targeted interventions. It also draws attention to the complexity and multiplicity of factors influencing human health and the realization that these factors cannot be effectively resolved piece mealy, but instead, must be approached from a systemic perspective. The focus of a systems perspective on a broad range of factors acting in tandem at various levels shifts the responsibility for health away from the individual and places it within the broader constraints of environmental, social and political forces under which the individual lives (Minkler 1989).

Third, the ecohealth approach recognizes that ecosystem change and the resulting health impacts do not fall evenly on society. Vulnerability and environmental health risks are shaped by a number of factors. For example, women and men may interact with their surrounding environment in specific ways that differentially shape their exposures to various environmental health risks. Examining issues from such a lens requires sensitivity to how different societal groups experience and respond to ecosystem change and the associated human health implications. A critical approach to ecohealth pays particular attention to how unequal power relations in society constrain peoples' access to, and use of ecosystem services and resources, and adversely impacts their health. For example, the approach explores how gender roles and relations in the household and in communities influence people's relationship with the environment and how such relations shape their health and well-being. However, as will be discussed in subsequent chapters, some critical scholars caution against focusing specifically on gender as a category, so as not to essentialize the various experiences. Instead they argue for the contextualization of men and women's experiences in relation to their multiple and changing identities and roles.

Finally, the ecohealth approach seeks to develop and implement interventions that transcend disciplinary boundaries, integrate knowledge across a variety of perspectives, and move health improvement beyond the boundaries of the health sector. The development of ecohealth interventions is a transdisciplinary and collaborative process. The ecohealth approach emphasizes the integration of various knowledge systems from complementary disciplines, including the natural, social, and health sciences. Professionals from these disciplines work with local actors and relevant stakeholders to collectively assess the environmental health problem at hand and develop appropriate interventions. This collaboration provides an opportunity for a transdisciplinary understanding of the problem, and allows for the development of solutions that are likely to be well-received by all stakeholders.

3.6 The Ecohealth Research Framework

The ecohealth approach serves as a useful framework from which to conduct research on issues at the interface of environment and health. It can also serve as a good framework for planning healthy communities or assisting communities to investigate and respond to an environmental health problem. Hence, there are a number of entry points from which to conduct an ecohealth research.

One entry point could be from a community perspective. A community might be facing a particular health concern that has a strong environmental component, but is uncertain about the linkages and causal pathways between the environmental conditions, their activities and the existing health conditions. For example, in certain regions in Canada, many First Nations and Aboriginal communities face issues of environmental contamination resulting from industry-related activities, such as the case of Grassy Narrows First Nations in Northern Ontario.[1] Between an 8-year period, the community fresh water was contaminated with up to 20,000 pounds of mercury that had been dumped from a paper mill located 320 km upstream. Residents started experiencing a host of health problems, including twitches, dizziness, eye problems and severe birth defects. In such a case, the use of an ecosystem approach to health will be instrumental in bringing together a variety of expertise to work with First Nations to evaluate the problem. An ecohealth research with a transdisciplinary team will allow for a concurrent investigation of a number of issues including, how the flora and fauna of the waterbody are affected, the causal

---

[1] Grassy Narrows protests mercury poisoning. Canadian Broadcasting Corporation. http://www.cbc.ca/canada/toronto/story/2010/04/07/tor-grassy-narrows.html. Accessed April 10th, 2010.

pathways through which mercury impacts human health, local people's understanding of their health experiences and the pathways through which their water and food sources are contaminated, as well as, the social, cultural, political, dimensions of the environmental problem. Similarly, a broad range of interventions may be proposed, ranging from individual precautionary measures to appropriate watershed remediation practices. These interventions may be cross-sectoral, and involve a variety of levels of government and departments. The private sector, in this case, is an important player, and needs to be actively involved in the research process.

Alternatively, a community may just want to plan to become a healthy community. Such a study will take into account existing environment and health problems, assess how these are linked to current and past community health and environment problems, assess the knowledge gaps, barriers and challenges toward achieving a healthy community, and identify and implement the necessary actions leading towards that future. A transdisciplinary approach will ensure that all dimensions of a healthy community are reflected in the community plan. Chapter 7 discusses an ecohealth approach to planning a healthy community in a West African village.

Two other entry points for ecohealth research are: (1) the identification of an environmental problem that is suspected to adversely impact human health; or (2) a well-documented health concern that is suspected to be associated with environmental degradation (Forget and Lebel 2001). While it is relatively easy to delimit the first scenario to a specific ecosystem such as a watershed, community, or mining site, it is difficult to confine the second scenario to a limited space (ibid). However, in both scenarios, it is important to make use of a transdisciplinary team of researchers, with the lead researchers having a good understanding of environmental phenomena in the first scenario, and health in the second , and the rest of the team comprised of experts from a variety of disciplines, including the natural, social, and health sciences, local people being impacted by the problem at hand, and other relevant stakeholders.

While ecohealth research makes use of external researchers, it is not a top-down approach. Instead, it makes use participatory procedures to work with the people experiencing the environmental health problem. Such research is geared towards generating new knowledge and understanding about the problem, while building research capacity and skills among the beneficiaries of the study. Most ecohealth research objectives are not amenable to traditional social research approaches, whereby an external researcher comes into a community, interviews the residents about the problem and single-handedly prescribes appropriate interventions, instead encourages the use of research approaches such as participatory action research (PAR). Participatory action research, which is described in detail in the next chapter has been used to conduct ecohealth research with communities in various regions, including Ghana and New Zealand (Dakubo 2004; Parkes and Panelli 2001)

There are five key elements central to conducting an ecohealth research. These include: (1) the integration of transdiciplinary, indigenous and stakeholder perspectives, (2) the use of participatory and inclusive procedures, (3) sensitivity to social diversities and their respective experiences and response to environmental health problems, (4) historizing environment and health problems, and (5) making use of critical perspectives to illuminate the political and social dimensions of human-environment relationships.

3.6.1 Integrating Transdisciplinary, Indigenous, and Stakeholder Perspectives

As a research framework, the ecohealth approach brings together researchers from a variety of disciplinary backgrounds, including sociology, anthropology, medicine, veterinary sciences, toxicology, ecology, among others. The integration of such disciplinary insights, together with those of traditional knowledge systems, allows for a comprehensive understanding of the problem to be investigated. The integration of different disciplines results in the emergence of unique and novel ideas that would not otherwise have emerged from a uni-disciplinary investigation. Peden (1999:3) describes transdisciplinarity as:

> ... going beyond disciplinary mind-sets into a re-conceptualization of phenomena, problems, goals, and approaches. [Transdisciplinarity] accepts complexity and pays attention to dynamic interactions (in space and time) between natural and human-made systems.

Local knowledge is an important component of transdisciplinary processes. Local people possess useful knowledge on how they interact with the biophysical environment, as well as unique perspectives of the structure and functioning of ecosystems. Inhabitants in many rural and Indigenous communities tend to have a close relationship with the natural environment. This close interaction endows them with in-depth understanding of ecosystem functions and processes. Also, local knowledges are useful in providing historical accounts of changes in ecosystem structure over time. As such local knowledge can contribute to understanding certain dimensions of ecosystem change that might not be readily apparent through scientific investigations. Despite the usefulness of local knowledge systems, researchers still have to be cautious in uncritically accepting local knowledges as accurate, current, or complete. As much as possible local knowledges should always be evaluated within the context in which they are produced. Much as local people are encouraged to participate in all aspects of the research process, it is impractical to expect local people to understand technical components of the research such as comprehending the biomolecular pathways through which mercury poisoning impacts the nervous system. Professionals will have to work with local people to come to a good understanding of such pathways.

Implementing transdisciplinary investigations can be fraught with many challenges given that many scientific investigators are accustomed to unidisciplinary procedures. However, when transdisciplinarity is achieved, and infused with local knowledge systems, it can allow for innovative solutions and new theories to emerge; giving rise to solutions that have increased chances of being successful than when developed from a single disciplinary perspective.

3.6.2 Making use of Collaborative and Inclusive Processes

Another important element of ecohealth research is the use of participatory proce-dures in working with local actors and relevant stakeholders to gather their views and insights about the environmental health issue under investigation. Participatory action research (PAR) is particularly suitable for conducting ecohealth research, given the intent to generate knowledge that can be put to action to solve a problem. PAR is a structured process of inquiry whereby those experiencing the problem collaborate with the researcher(s) as partners to identify and address the prob-lem through a process of research, action and reflection (Fals-Borda and Rahman 1991; Whyte 1991). Usually, those experiencing the problem are not just pas-sive victims, but also active players in finding solutions to the problem. They participate in all stages of the research process, including the delineation of the problem, the collection of relevant data, and the analysis and interpretation of such data. Through active involvement, beneficiaries of the study become com-mitted to the research outcome and are more likely to implement or adopt the proposed solutions. However, achieving true participation can be a tricky exercise since not all issues are amenable to group or public processes. Hence participatory approaches should be complemented with follow-up individual interviews so as to obtain relevant information that would otherwise not be captured through partici-patory processes. Also power dynamics in groups can result in gathering dominant views that may not necessarily be representative of the issue under investigation. Similarly, the unequal power relationships between local actors and scientific inves-tigators can inhibit true collaboration and equal partnerships that is often required in many PAR projects. Despite these challenges the use of participatory action research encourages meaningful participation by all stakeholders in all phases of the research process resulting in increased learning and understanding of the problem at hand.

3.6.3 Taking Heterogeineity and Difference into Account

The ecohealth approach recognizes that due to unequal power relations in society, certain groups of individuals or regions in the world will become more vulnerable to the health impacts of ecosystem change than others. Environmental costs in the form of pollution or degraded natural resource base are likely to be displaced to these vul-nerable groups, who more than likely will be unable to cope with the health effects of these environmental costs. The ecohealth approach also recognizes that due to social, cultural, economic and political factors, people interact with the ecosystem in different ways. These varied interactions are likely to expose people to different health risks. These varied experiences, identities, and roles must be factored into any ecohealth research project. However, care must be taken not to essentialize certain identities, roles or experiences, as these are dynamic and should always be examined and interpreted based on context.

3.6.4 Historicizing Environment and Health Problems

Historical information plays an important role in understanding or interpreting the causal factors and processes responsible for ecosystem degradation or the persistence of certain health conditions in various communities or population groups. For example, many current health and environmental practices and policies in the developing world, especially in Africa, still have their roots in the colonial context, and the failure to acknowledge these historical antecedents may result in a wrongful analysis or interpretation of a particular environmental or health phenomenon. For example, current social patterning in health outcomes between Northern and Southern Ghana, and between rural and urban areas in Ghana can be explained based on colonial powers decision to locate health care services and facilities in coastal urban centres where colonial masters resided, to the detriment of northern and rural areas of the country. Similarly, the preference for curative versus preventive approaches to health care can be traced to colonial health policies, whereby treatment was the preferred mode of care so as to allow for a healthy labour force to continue with mineral extraction and other capitalist expansion activities. Preventive public health measures such as the provision of clean water, adequate sanitation, and proper housing were considered too expensive and so were restricted only to colonial masters (Randall 1998). Even though many African countries have gained independence, many policies are still externally determined and do not differ much from the colonial ones. So, it is important for an ecohealth research project to take into account these historical contexts as they allow for a better understanding of certain health practices, behaviours and policies.

Similarly in the environmental sector, colonial policies were used to designate protected areas, such as forest reserves and game reserves. These protected areas were restricted for use by colonial masters, and not necessarily for ecological reasons. Today, some of these protected areas still exist, but the reasons have changed to environmental conservation. Because these ecosystems are the primary source of livelihood for many local communities, poaching, illegal farming, fuel wood harvesting and encroachment take place in these designated spaces. Without understanding such historical contexts, these practices could be misconstrued as deliberate acts of ecosystem distruction, and constitute a starting point for an ecohealth research project. Ecohealth research initiatives must therefore investigate the historical antecedents of environmental degradation in order not to mislabel or misdiagnose an environmental problem. A misdiagnosed problem means a wrong solution, and ecosystem degradation will continue to persist.

3.6.5 Infusing Critical Social Theory into Ecohealth Research

With few exceptions, most ecohealth research has not engaged with critical theoretical developments such as poststructuralist political ecology, discourse theory, the sociology of science and the politics of environmental and public health knowledge claims. The use of such theoretical perspectives allows for a comprehensive and critical understanding of the underlying and "real" causes of ecosystem degradation and the forces shaping poor health. Most of the literature still attribute the major driving factors of ecosystem degradation to deforestation, desertification, poverty, and rapid population increases without examining the underlying drivers, including the political constructions of these so-called driving factors. For example, a political ecology approach to ecohealth research illustrates the centrality of politics and power dynamics in shaping human-environment interactions and how such power dynamics contribute to the uneven production and distribution of health risks (Farmer 2001). The application of poststructuralist perspectives to political ecology of health further allows for an interrogation of the taken-for-granted assumptions and causal explanations of ecosystem degradation and ill health. Critical theoretical perspectives draw attention to how social and political framings are incorporated into scientific knowledge claims and how policies and interventions emanating from such "un-reconstructed" science (Forsyth 2003) might further increase environmental degradation or ill health. This book builds on these arguments and articulates a pathway for incorporating such critical theoretical perspectives into ecohealth research.

3.7 Conclusion

The ecosystem approach to human health is an emerging field of study, and the key attributes of this approach are continuously evolving and benefiting from new theoretical developments. However, at the core of the ecohealth approach are the concepts of human health and ecosystem health. The concept of an ecosystem serves a dual purpose of emphasizing the important role of ecological factors in influencing human health, as well as provides a conceptual framework from which to examine the interactions of the multitude of factors shaping human health. Also, as a research framework, the ecohealth approach emphasizes the use of transdisciplinary, participatory, and inclusive processes. The goal is to work with all relevant stakeholders to better understand how human activities transform the ecosystem and adversely impact our health. It also seeks interventions that aim to improve both human health and ecosystem health with the increasing recognition that an unhealthy environment cannot sustain a healthy population.

References

Bell A (1994) Non-human nature and the ecosystem approach: the limits of anthropocentrism in Great Lakes management. Alternatives 20:20–25

Costanza R (1992) Toward an operational definition of ecosystem health. In: Costanza R, Norton BG, Haskell BD (eds) Ecosystem health: new goals for environmental management. Island press, pp 239–256

Costanza R, Norton BG, Haskell BD (1992) Ecosystem health: new goals for environmental management. Island Press, Washington, DC

Dakubo C (2004) Ecosystem approach to community health planning in Ghana. EcoHealth 1: 50–59

Duhl LJ (1976) The process of recreation: the health of the "I" and the "Us". Ethics Sci Med 3:33–63

Fals-Borda O, Rahman MA (1991) Action and knowledge: breaking the monopoly with participatory action research. Apex Press; Intermediate Technology Publications, New York, NY

Farmer P (2001) Infections and inequalities: the modern plagues. Updated Edition. University of California Press, Berkeley, CA

Forget G, Lebel J (2001) An ecosystem approach to human health. Int J Occup Environ Health 7(2 Suppl):S3–S38

Forsyth T (2003) Critical political ecology: The politics of environmental science. Routledge, London

Fox NJ (1991) Postmodernism, rationality and the evaluation of health care. Sociol Rev 39(4): 709–744

Fox NJ (1994) Postmodernism, sociology and health. University of Toronto Press, Toronto, ON

Hancock T (1990) Towards healthy and sustainable communities: health, environment and economy at the local level. A presentation at the 3rd colloquium on environmental health. Quebec, Canada

International Development Research Centre (1997) An approach and method for assessing human and environmental conditions and progress toward sustainability: overview. IUCN-The World Conservation Union. IDRC Assessment Tools. Ottawa, ON

Karr JR (1991) Biological integrity: a long neglected aspects of water resource management. Ecol Appl 1:66–84

Karr JR, Fausch KD, Angermeier PL, Yant PR, Schlosser IJ (1986) Assessing biological integrity in running waters: a method and its rationale. Ill Nat Hist Surv Spec Pub 5 Champaign

Kay JJ (1993) On the nature of ecological integrity: some closing comments. In: Woodley S, Kay, J, Francis G (eds) Ecological integrity and the management of ecosystems.St. Lucie Press, Delray Beach, FL

Kay J, Regier HA, Francis M, Francis G (1999) An ecosystem approach for sustainability: addressing the challenge of complexity. Futures 31:721–742

Kerr SR, Dickie LM (1984) Measuring the health of aquatic ecosystems. In: Cairns VW, Hodson PV, Nriagu JO (eds) Contaminant effects on fisheries. Wiley, New York, NY, pp 279–28

Labonté R (1991) Econology: integrating health and sustainable. Part one: theory and background. Health Promot Int 6:49–65

Lebel J (2003) In focus: an ecosystem approach. IDRC, Ottawa, ON

Lincoln Y (1992) Fourth generation evaluation, the paradigm revolution and health promotion. Can J Public Health 83:S6–S10

Maini JS (1992) Sustainable development of forests. Unasylva 43:3–8

Millennium Ecosystem Assessment Series (2005). Ecosystems and human well-being: a Framework for Assessment; Ecosystems and Human well-being. Available online at: http://www.millenniumassessment.org/. Accessed 10 May 2010

Minkler M (1989) Health education, health promotion and the open society: An historical perspective. Health Educ Behav 16:17–30

Nielsen NO (2001). Ecosystem approaches to human health. Cad Saúde Pública 17: Suppl:69–75

Noack H (1987) Concepts of health and health promotion. In: Abelin, T etal (eds) Measurement in health promotion and protection. WHO, Copenhagen, pp 5–28

Odum EP (1953) Fundamentals of ecology. W. B. Saunders, Philadelphia, PA

Odum EP (1985) Trends expected in stressed ecosystems. Bioscience 35:419–422

Parkes M, Panelli R (2001) Integrating catchment ecosystems and community health: the value of participatory action research. Ecosyst Health 7:85–106

Peden DG (1999) Mono-, multi-, inter-, and trans disciplinarity in IDRC research activities. International Development Research Center, Ottawa, ON

Randall P (1998). Health care systems in Africa: Patterns and Prospects. Report from the workshop, Health Systems and health care: Patterns and Perspectives. 27–29 April 1998. The North-South Co-ordination Group. University of Copenhagan and The ENRCA Health Network

Rapport DJ (1992) Evaluating ecosystem health. J Aquat Ecosyst Health 1:15–24

Rapport DJ (1995). Ecosystem health: an emerging, integrating science. In: Rapport DJ, Gaudet CL, Callow P (eds) Evaluating and montinoring the health of large-scale ecosystems. NATO ASI Series 1, (28). Springer, Germany, pp 5–31

Rapport DJ et al (2001) Ecosystem health: the concept, the ISEH, and the important tasks ahead. Ecosyst Health 5:82–90

Rapport DJ, Costanza R, McMichael AJ (1998) Assessing ecosystem health. Trends Ecol Evol 13:397–402

Schindler DW (1990) Experimental purtabations of whole lakes as tests of hypotheses concerning ecosystem structure and function. Oikos 57:25–41

Seedhouse D (1986) Health: the foundations of achievement. Wiley, Chichester

Steedman RS, Regier HA (1990). Ecological basis for an understanding of ecosystem integrity in the Great Lakes Basin. Proceedings of a workshop on integrity and surprise, June 14–16, 1988. International Joint Commission, Windsor, Ontario and Great Lakes Fishery Commission, Ann Arbor, MI

Tansley AG (1935) The use and misuse of vegetational terms and concepts. Ecology 16:284–307

Van Leeuwen J, Waltner-Toews D, Abernathy T, Smith B (1999) Evolving models of human health toward an ecosystem context. Ecosyst Health 5:204–219

Waltner-Toews D (2001) An ecosystem approach to health and its applications to tropical and emerging diseases. Cad Saude Publica 17(Suppl:7–22):discussion 23–36

Whyte WF (1991) Participatory action research. Sage, Newbury Park, CA

World Health Organization (1948) Preamble to the Constitution of the World Health Organization as adopted by the International Health Conference, New York, 19–22 June, 1946; signed on 22 July 1946 by the representatives of 61 States (Official Records of the World Health Organization, no. 2, p. 100) and entered into force on 7 April 1948.

World Health Organization (1978) Alma-Ata 1978. Primary health care, Geneva

World Health Organization (1986) Health promotion: Ottawa charter. International Conference on Health Promotion, Ottawa, 17–21 November 1986. Geneva, Switzerland

Part II
Methodological Approaches and Processes for Conducting Ecohealth Research

Chapter 4
Community-Based Participatory Research for Ecohealth

Contents

4.1 Introduction

Community-based participatory research (CBPR) is particularly suitable for conducting ecohealth research. As discussed in the previous chapter, one primary objective of ecohealth research is to create opportunities for all relevant stakeholders to participate in evaluating the environmental health problem at hand, and to gain sufficient insight to implement the appropriate interventions. Community-based participatory research in public health is an investigative approach that actively involves all stakeholders, including community members, representatives from various institutions, and researchers, in all phases of the research process, to evaluate and respond to particular health concerns in a proactive manner (Israel et al. 2003; Laverack and Labonte 2000; Robertson and Minkler 1994; Schwab and Syme 1997). The goal is to bring together the unique perspectives, experiences and knowledge systems about a particular problem, and integrate the knowledge gained with action to improve the health and well-being of community members (Israel et al. 2003). Community-based participatory research is very suitable to ecohealth research since the ecohealth approach calls for the integration of perspectives from various disciplines with those of relevant stakeholders.

C.Y. Dakubo, *Ecosystems and Human Health*, DOI 10.1007/978-1-4419-0206-1_4,
© Springer Science+Business Media, LLC 2011

Compared to traditional applied social science approaches to health research where the researcher, mostly an outside expert, dominates the research process and single-handedly identifies and investigates a particular health problem, CBPR seeks to generate learning through participation and action, with the ultimate goal of giving voice and power to those affected by the problem to influence the research process and outcome (Fals-Borda and Rahman 1991; Gaventa 1981, 1993; Hall 1992; Maguire 1996; Wallerstein 1999). CBPR confirms the value of communities' experiential knowledge by working with people to strengthen their awareness of their own capabilities both as researchers and change agents (Hagey 1997). CBPR consciously makes use of methods that blur the lines between the researcher and the researched (Gaventa 1981). By doing so, CBPR shifts the decision-making authority away from experts and professional researchers to embrace the experiential knowledge of the ordinary person and other stakeholders (Brooks and Watkins 1994).

Community-based participatory research in public health emerged, partly in response to the growing inability of traditional approaches to health research to effectively address the growing complexities and contextual factors shaping public health outcomes, as well as explain the growing health disparities between various regions and among population groups. For example, why does poor health continue to persist in developing countries, in Indigenous communities, in rural areas, despite years of public health interventions? How is it that these groups continue to bear a disproportion portion of the global burden of disease? What role and opportunities do these groups have in contributing to their own health and well-being? The answers to these questions are complex, and are not amenable to traditional top-down approaches to health research, that focus on collecting data on individual level risk factors, with little consideration on how socioeconomic, political and environmental factors contribute to shaping people's health outcomes and disparities. Besides, newly emerging diseases such as HIV/AIDS, SARS, Ebola, among others defy conventional, individual-focused public health approaches, instead requiring the integration of expertise that crosses disciplinary boundaries, and blurs the boundaries between lay and expert opinions.

CBPR tends to be used as an umbrella term to describe a variety of participatory research approaches, such as participatory action research (PAR), feminist participatory research, action learning, among others. Although these approaches vary in their goals and change strategies, they share a set of core assumptions, features, values and principles. For example, almost all facets of CBPR engender a participatory, co-operative, and co-learning process. They encourage a process that develops and builds local capacity and empowers people to increase control over their lives. CBPR also seeks a balance between research and action (Israel et al. 1998; Minkler and Wallerstein 2003; Wallerstein 1999).

In addition, CBPR approaches are thought to comprise of three major components: participatory research, education, and social action (Leung et al. 2004). The *participatory research* component provides opportunities for beneficiaries of the study to participate in collectively analysing the problems facing their community and become active players in finding solutions to those problems. By participating in such a collaborative research processes, participants take equal ownership of the research process, ensuring that the research outcomes are practical, implementable, and relevant to their specific interests and needs. Education is an essential component of participatory research processes. Through participation in the research process, participants acquire knowledge that is use to the situation at hand, and can be used to solve similar problems in the future. Participants also acquire new research-related skills that can be used to investigate community problems, including environmental health problems. Through education, participants engage in discussions that create critical awareness and consciousness about their problems and how these relate to the broader social structure (Yeich and Levine 1992). The third component of CBPR is social action. Social action is one that distinguishes CBPR from traditional research approaches. Most often than not, traditional research generates findings that end up in scholarly publications, with the implications of the study rarely translated to inform social action or change. In contrast, CBPR views action as an integral component of the research process, and encourages the identification of practical actions that can be implemented to address the problem at hand. The actions to be undertaken are usually jointly agreed upon by the beneficiaries of the study and their research partners (Leung et al. 2004). Participatory action research (PAR) is a common CBPR approach that has been used in a number of ecohealth research projects, including some of the case studies presented in subsequent chapters. To that end, it is important that we examine some of the key principles of PAR and how this is used to conduct ecohealth research.

4.2 Participatory Action Research

Participatory action research (PAR), sometimes used interchangeably with participatory research, is a structured process of inquiry in which those experiencing a problem collaborate with the researcher(s) as partners to identify and respond to the problem through a process of research, action and reflection (Fals-Borda and Rahman 1991; Israel et al. 1998; McTaggart 1991; Parkes and Panelli 2001; Whyte 1991). PAR emerged among a group of research alternatives aimed at responding to the failure of conventional applied social science methods to understand and address complex social problems and implement successful interventions (Brown and Tandon 1983; Corcega 1992; Maguire 1987; Park et al. 1993). The roots of PAR, especially its action and participation components, are usually traced to two traditions: (1) the "action research" school developed by Kurt Lewin (1946) in the 1940s, and (2) the emancipatory research school developed by Paulo Freire (Freire 1982) and other third world scholars in the 1970s (Park et al. 1993; Tandon 1996). In the action research tradition, which is sometimes referred to as the Northern tradition, those experiencing a problem are involved in a cyclical, iterative process of problem identification, fact finding, analysis, implementation and monitoring. Lessons from the monitoring stage are fed back to problem identification (Hart 1996; McTaggart 1991, 1997; Reason 1994). Through this iterative process, action research does not only engage in research and action, but also generates learning as part of the problem solving process (Peters and Robinson 1984).

The second tradition, sometimes called, the Southern tradition, emerged through Paulo Freire's emancipatory research, which attempted to counter the "colonizing" nature of the research that was being conducted *on* oppressed people in the South, particularly Latin America, Asia, and Africa (Brown and Tandon 1983; Fals-Borda and Rahman 1991; Freire 1982). Freire's tradition considers collective participation as critical to challenging inequities and ensuring social progress. Freire's philosophy is that when people engage in dialogue about their communities and social conditions surrounding them, they become more aware and knowledgeable about the problems facing them and can then begin to chart a path for solutions them. Also, the bond between people becomes strengthened, and people become capable in their ability to reflect on their own values and choices. Freire (1982) saw this process of conscientization and education as a path to human liberation, which means that people become active players in their own learning, and not passive empty vessels waiting to be filled with expert knowledge. The PAR process therefore embodies these qualities of participation, education, action and social change. PAR researchers have the duty then to facilitate this process of learning by creating a research environment in which participants can take greater control of the research process, and become committed to implementing the necessary actions for social change (Cornwall and Jewkes 1995; Hagey 1997; Hall et al. 1982; Maguire 1987).

PAR differs from other participatory approaches in that it focuses on both process and product objectives. The process in a PAR research is equally as important as the outcome of the study. It is through the process that knowledge is acquired, and ownership, control, and commitment to the outcome all become enhanced. Unlike conventional applied social science research approaches that sometimes investigates a topic that is of interest only to the researcher, PAR tends to investigate research topics that are of interest both to the community or people experiencing the problem and the researcher, and usually aims to combine the knowledge and expertise of both players to reach a mutually acceptable course of action, which could be to improve community health and eliminate health disparities (Minkler and Wallerstein 2003). Similar to most CBPR procedures outlined above, the PAR researcher and stakeholders engage in a collaborative, joint process of inquiry; together they decide the focus of knowledge generation, collect and analyse data, and take action to solve the problem at hand. Through this collaborative investigation and reflective dialogue, community members and other stakeholders learn to critically analyse their own problems and devise solutions to them. This process offers an educational experience that serves to respond to community needs and motivate people to implement the solutions developed.

Participatory action research has been widely used in development programs and in fields such as agriculture, community development, adult education, and community-based natural resource management projects. However, in the past few decades, participatory research approaches are increasingly being used in the field of public health, partly because of the inability of the biomedical approach to effectively respond to the cultural, social, ecological, and political dimensions of health.

Improving human health requires understanding peoples' beliefs, attitudes and behaviors, and working with them to understand these viewpoints and incorporate them into the research process and identify appropriate interventions.

Within the context of ecohealth, a PAR project brings together researchers from various disciplines, community members, and key stakeholders to collectively investigate environment and health issues, and develop appropriate measures. Environment and health problems can benefit from both the expertise of professionals and local actors. Bringing such knowledge systems together through participatory action research generates new learning about ecosystem conditions, the pathways through which ecosystem degradation adversely impacts human health, and an exploration of feasible actions to improve both human health and ecosystem health (Dakubo 2004).

4.3 Being Critical About Participatory Research Approaches

While participatory research approaches have gained prominence among academics over the years, they have also come under extensive criticisms. For example, participatory action research has been criticised internally by its users as well as externally. External critics are concerned of the lack of clear distinction of PAR from other approaches such as community organizing, organizational development, among others (Greenwood 1994; Hart 1996). Critics argue that the extent to which the many variants of definitions of PAR distinguish itself from similar approaches is not clear. In response, Greenwood (1994) argues that, while PAR shares characteristics with other approaches, it is distinct in its ability to combine the three components of research, education and action, and also in its ability to integrate and validate local knowledges in the research process. PAR has also been criticized for not adequately distinguishing between research and practice, in other words blurring the lines between the two areas (Hitchcock and Hughes 1995; Meyer 1993). However as Fine (1994: 30) explains such blurring of lines within research projects helps the researcher to negotiate the "messy nexus of theory, research and organizing". This lack of clear distinction also reinforces the need for reflexivity through the research process, and constantly being aware of our positionality as researchers, and who and what we represent (Reason 1994).

From an internal perspective, the criticisms have focused on the mutual identification of a research problem that is supposed to be of interest to both the researcher and the community. It is very unlikely that a research problem can mutually satisfy the career interests of an academic researcher as well as align perfectly with community interests, wants and needs. It is realistic to expect a balance to be reached. The are also questions about who initiates the process, how, where, and when (Eisen 1994; Wallerstein and Bernstein 1994). The concern here is that many communities in need, especially rural and remote communities, lack that inertia and initial drive to initiate a PAR project. Also many of these communities are too busy going about their daily lives that, although they might be aware of a problem, they lack the resources, interest, and commitment to seek a solution. So, most often than not, it is the external researcher who normally has the time and resources to initiate and facilitate a PAR project. However, in such circumstances, and along the requirements of PAR, the researcher must involve community members to mutually share the concern that the issue to be investigated is a shared concern, otherwise the researcher will be imposing an external agenda on the community, which might not necessarily reflect their concerns.

Another set of criticisms comes from critical scholars who are concerned about the uncritical use of concepts such as community, participation, emancipation and empowerment in participatory research approaches (Cameron and Gibson 2005; Kothari 2001). Such critics argue that, while participatory research draws on post-positivist approaches, it fails to pay attention to local power differences by assuming a common community perspective; or the existence of "marginalized", "oppressed", and "silenced" subject positions. They also argue that PAR overly emphasizes personal reform over political struggle, and uncritically employs a language of emancipation to incorporate and change existing conditions (Williams 2004). The uncritical use of these concepts may lead to the production of constraining, rather than liberating knowledges and policies. Also care must be taken in seeking consensus, as this reduces different, diverse, and unique experiences into coherent and homogenous experiences, while preventing the micro-level struggles and inequities to emerge. Below we examine these concepts from a critical perspective.

4.3.1 Being Critical About "Community"

Within the new public health discourse, community participation is seen as a key feature that separates it from the individualistic ethic of the biomedical approach. Community participation is seen as a means to provide people with the opportunity to identify and self-define their needs and to participate in a thoughtful reflection as to how to contribute to solving these problems (Hawe 1994; McKnight 1987). Zakus and Lysack (1998: 2) define community participation as the process by which members of a community, either individually or collectively, and with different levels of commitment:

(1) develop the capability to assume greater responsibility for assessing their needs and problems;
(2) plan and then act to implement their solutions;
(3) create and maintain organizations in support of these efforts; and
(4) evaluate the effects and bring about necessary adjustments in goals and programs on an ongoing basis.

Despite the emphasis of involving community members in health promotion strategies, the concept "community" and how it is used still remains contested in the literature. For example, in the health literature, although there is no formal agreement on how community should be defined, an implicit definition can be derived from how community is used in various documents such as the Ottawa Charter for Health Promotion and the Alma-Ata Declaration on Primary Health. The Alma-Ata Declaration views community as a place-based aggregation of people with commonly shared social, economic, political, and cultural needs (1978). Other conceptions present community as a coherent unit, whose members work towards commonly shared goals (Jewkes and Murcott 1996); or a social system, in which the focus is on social interaction, social institutions and social control (Haglund et al. 1990); or a "community of interests" which represents a group of people sharing norms and values, such as Indigenous communities.

Most of these conceive community as a spatial or social unit, with distinctions such as rural and urban, micro and macro, and local and global (Dickens 1990). They also see community as having people with commonly shared goals, priorities, and dreams. Rather than focus on such dualisms and commonalities, critical theorists emphasize the need to contextualize communities based on the multitude pressures that are coming their way and how they are adapting to these pressures. As Nilsen observes, communities should be conceived based on "how external influences are shaping the defence, reconstructions or constructions of local identities, and creating new potentials for collective action and local enthusiasm" (Nilsen 1996: 170).

In addition, there is also caution in assuming pre-existing categories of community. Communities do not exist a priori, they are constructed, defined, and labelled by various discourses and practices (e.g. a deserted community, a "ghost" town). We define them according to our needs and the issue to be investigated, and so they can be defined geographically, by the primary resource base, or by other means. Similarly, within communities, there are no pre-existing identities of community members such as "oppressed", "marginalized", instead community members should be seen as always in the process of being constructed by various discourses, actions and practices (Foucault 1979; Cameron and Gibson 2005). Thus, when we go to the field, we should not associate a single identity with a "community" or its membership. Instead, community membership should be seen as a dynamic relation of power structures, with varied and changing representations, needs, views and knowledges. Hence people's views about the environment and about health problems should be contextualized based on their respective circumstances and identities, including gender, social status, educational level, among others. Critical perspectives also caution against the uncritical adoption of local views as representative of the community, without situating it within the respective circumstances of the provider.

In the context of ecohealth research then, it is important to refrain from approaching the "community" "site" or "village" as a spatially-bound entity with clear and uncontested membership, or perceive it based on the environment or health problem at hand,– a mining town, a coastal town, or a farming community. Such pre-defined notions of a community could be limiting, historical in scope, and probably have nothing to do with the current needs and opportunities of the community. Besides community members could be bound by social and cultural norms that supersede how they are perceived externally. Conducting ecohealth research with an uncritical notion of "community" risks masking existing conflicts and local power structures, smoothening over repressive structures that operate at varying levels, including the micro-level (e.g. gender, class, and ethnicity) and the global level (implications of unfair trade agreements); and risk celebrating community/local knowledges as authentic and representative of the community (Kothari 2001; Mohan 2001).

4.3.2 Being Critical about "Participation"

Like community, the concept "participation" has proven difficult to define, initiate and sustain in participatory research processes. The primary objective of most participatory research projects is to involve beneficiaries in various aspects of the research, ideally, all phases of the research. The processes used to actively engage people and the extent to which people are engaged in the research process are still unresolved issues. According to Rifkin (1996), the extent to which people are involved in the research process vary along a continuum from manipulation or tokenistic forms of participation, in which the researcher attempts to get community members to own an externally defined research agenda, all the way to levels of full community participation. Full community participation occurs when the researcher forms an equal partnership with community members, recruits them as co-researchers and together they identify solutions to the problem. At the full participation level, community members participate in all phases of the research process, including the identification of research questions, project design, data collection and analysis, and interpretation of findings. Such equal partnership ensures that the problems are accurately diagnosed and the results collectively owned. The commitment to implement solutions or actions coming out of such research projects is enhanced through such partnership efforts.

How participation is deployed is also important in determining the extent to which it achieves its objective of active engagement. Parfitt (2004) suggests that it is important to first understand whether or not participation is deployed as a means or as an end in itself. Parfitt argues that to the extent that participation is a means, it will be difficult achieving equal partnerships between the researcher and community members. In such situations, community members are mobilized to participate in an externally-driven research agenda. Participation as a means takes a relatively short period of time and puts little emphasis on understanding and critically analysing "the community" (Oakley et al. 1991). In such research processes, the researcher (mostly an outsider) dominates the entire research process and occasionally seeks information from some key people in the community. This approach, sometimes called the "expert" approach, is suitable for problems that require expert knowledge and advice, for example addressing oil spills in a community.

In situations where participation is an end in itself, the goal is to blur the unequal power relations that exist between the professional or academic researcher and beneficiaries of the study, and to create a space for active engagement by all relevant stakeholders. The goal usually is to generate learning and empower participants to respond proactively to the problem. In such circumstances, it is important to understand how people are chosen to participate, why, and the extent to which they participate in finding solutions to the problem pay attention to questions concerning who participates, how, why, and the extent (Cornwall and Jewkes 1995; Rifkins 1996).

Within the context of ecohealth research, then the goal is to actively engage community members in all phases of the research with attention being paid to both the process and the outcome. Because transdisciplinarity is a key component of the ecohealth research process, issues that are technical in scope will benefit from the expertise and insights of professional researchers in the team. It is incumbent on members of the transdisciplinary research team to try as much as possible to explain in very lay terms to other stakeholders any technical issues that might be part of the investigation, so as to actively keep them involved and interested in the process. Until such efforts are made, many participatory research projects will always fall short of true or full participation. Also because ecohealth research usually brings together professionals from various disciplinary backgrounds to work collaboratively with community members, it is often difficult building equal partnership among the professionals themselves, let alone between professionals and ordinary citizens. These challenges and how to respond to them are discussed in subsequent chapters.

4.3.3 Being Critical About "Empowerment"

Community empowerment is seen as a primary strategy for health promotion (Laverack and Labonte 2000; Labonte 1996; Wallerstein 2002). Like community participation, various scholars have interpreted the concept of "community empowerment" differently. Underlying all the variants is the agreement that empowerment is a process through which individuals, groups, communities and populations become more involved and are able to take control of and make decisions about their own health and well-being (Scriven and Stiddard 2003; Laverack and Labonte 2000). In the context of health promotion, this would entail people taking increased control over their personal health behaviour, or advocating for the basic necessities of health such as clean water, adequate housing, proper sanitation, among others (Laverack and Labonte 2000).

However, like participation, the means through which people are empowered are fraught with difficulties. According to Rappaport (1985), empowerment does not occur when power is given, instead empowerment takes place when that power is taken by individuals and communities themselves and used to help them identify and work towards achieving their own goals. The role of the researcher then is to provide the necessary conditions for empowerment to occur and to nurture the process by resisting the urge to lead or take responsibility (Labonte 1989: 87). The researcher plays the role of a facilitator or a coach in team-building by working with community members to articulate both their health problems and the solutions to address those problems. How then do we determine whether empowerment has been achieved? Indicators of true empowerment include attributes such as self-esteem, individual and community competence, self-confidence, self-awareness, self-development and improved quality of life (Wallerstein and Bernstein 1988). Also, there is the assumption that empowerment occurs when people are fully engaged in the research process. This is in line with Paulo Freire's (1972, 1973) principles of critical consciousness where he asserts that community empowerment begins to occur when people are able to engage in participatory dialogue, and are involved in naming their problems, and identifying new ways to transform oppressive structures.

However, related to earlier discussions on the ambiguities with "participation" and "collective community voices", Wallerstein (1992) cautions that in our attempt to empower, it is important to ask: Who exactly are we empowering? Does empowerment mean that some individuals or groups gain at the expense of others? Does empowerment sufficiently challenge power structures that systematically operate to marginalize others and leave them in poorer health than others? Wallerstein cautions that the assumption of a liberating and empowering role of participatory research could end up forcing participants to participate in their own oppression by assuming and championing the goals and priorities of elite groups in the community, while silencing theirs (Wallerstein 1992).

There are also false assumptions that by mobilizing people to participate in participatory research projects, the silenced voices of the marginalized will be released, thus enabling them to confront the structural barriers affecting them and place them on the path to emancipation, liberation and empowerment (de Roux 1991; Park 1993). However, related to earlier discussions about pre-existing identities, critical scholars argue that there are no pre-existing subject positions that are repressed and in need of empowerment or liberation. The identities of silenced voices, empowerment, and liberation are all subject positions that are created by the way we use language. Rather than assume the existence of pre-existing categories of repressed, silenced, etc., critical scholars argue that it is beneficial to see through the lens of various forms of subjection, of which empowerment and liberation are some (Cruikshank 1999).

Finally, there seem to be some concerns with empowerment as a means of transferring or imparting skills and analytical abilities to people, as this implies that professionals or researchers have superior skills and knowledge, over those of local people. This means valorizing some knowledge claims, while silencing others. Instead knowledge should be seen as acquired through the research process, and then internalized to raise consciousness and consequently empowerment.

4.4   Conclusion

Community-based participatory research proves to be a useful approach to conducting ecohealth research, as it provides the opportunity for community members and other stakeholders to participate in evaluating and solving the environment and health problems facing the community. It also provides an opportunity for members of the transdisciplinary research team to work collaboratively with local actors in all stages of the research process. Through this collective investigation, both professional researchers and local actors find ways to integrate their unique perspectives, thereby allowing for a good understanding of the problem at hand, and also allow for new strategies of responding to environment and health problems to emerge. However, as discussed above, CBPR is not without its challenges, and sometimes it is difficult to achieve that ideal level of participation or consciousness raising that is often desired. In addition, it is important to be critical about how we use terms such as community, participation, and empowerment, and always cognizant that these features are not static but always in flux, and as such must always be interpreted based on context. In the next chapter, we will discuss how to operationalize an ecohealth participatory action research project in the field, discussing the practical steps of setting up a transdisciplinary research team, recruiting participants, and conducting an ecohealth research.

References

Brooks A, Watkins KE (1994) The emerging power of action inquiry technologies. Jossey-Bass, San Francisco, CA

Brown J, Tandon R (1983) Ideology and political economy in inquiry: actionvresearch and participatory research. J Appl Behav Sci 19:277–294

Cameron J, Gibson K (2005) Participatory action research in a poststructuralist vein. Geoforum 36:315–331

Corcega TF (1992) Participatory research: getting the community involved in health development. Int Nurs Rev 39(6):185–188

Cornwall A, Jewkes R (1995) What is participatory research?. Soc Sci Med 41(12):1667–1676

Cruikshank B (1999) The will to empower: democratic citizens and other subjects. Cornell, Ithaca

Dakubo C (2004) Ecosystem approach to community health planning in Ghana. EcoHealth 1: 50–59

De Roux G (1991) Together against the computer: PAR and the struggle of Afro-colombiams for public services. In: Fals Borda O, Rahman MA(eds) Action and knowledge: breaking the monopoly with participatory action research. The Apex Press, New York, NY, pp 37–53

Dickens C (1990) Urban sociology. Society, locality and human nature. Harvester Wheatsheat, Hemstead

Eisen A (1994) Survey of neighborhood-based, comprehensive community empowerment initiatives. Health Educ Q 21(2):235–252

Fine M (1994) Distance and other stances: negotiations of power inside feminist research. In: GitlinA (ed.) Power and method: Political activism and educational research. Routledge, New York, NY, pp 13–35

Foucault M (1979) Discipline and punishment: the birth of the prison (trans: Sheridan A). Vintage, New York, NY

Freire P (1972) Pedagogy of the oppressed. Herder and Herder, New York, NY

Freire P (1973) Education for critical consciousness. Seabury Press, New York, NY

Freire P(1982). Freire P. Creating alternative research methods: learning to do it by doing it. In: Hall B, Gillette A, Tandon R (eds) Creating knowledge: a monopoly? Participatory research in development. Society for Participatory Research in Asia, New Delhi, pp 29–37

Fals-Borda O, Rahman MA (1991) Action and knowledge: breaking the monopoly with participatory action research. Apex Press; Intermediate Technology Publications, New York

Gaventa J (1981) Participatory action research in North America. Convergence 14:30–42

Gaventa J (1993) The powerful, the powerless, and the experts: knowledge struggles in an information age. In: Park P, Brydon-Miller M, Hall B, Jackson T (eds) (1993) Voices of change: participatory research in the United States and Canada. Bergin and Garvey, West Port, CT, pp 21–40

Greenwood J (1994) Action research: a few details, a caution and something new. J Adv Nurs 20:13–18

Hagey RS (1997) The use and abuse of participatory action research. Chronic Dis Can 18(1):1–4

Haglund BJA et al (1990) Assessing the community: its services, needs, leadership and readiness. In: BrachtN (ed) Health promotion at the community level. Sage, Newbury Park, CA

Hall BL (1992) From margins to center? The development and purpose of participatory research. Am Sociol 23:15–28

Hall, BL, Gillete, A, and Tandon, R (eds) (1982) Creating knowledge: a monopoly? Participatory research in development. Society for participatory research in Asia, New Delhi

Hart E (1996) Action research as a professionalizing strategy: issues and dilemmas. J Adv Nurs 23:454–461

Hawe P (1994) Capturing the meaning of 'community' in community intervention evaluation: Some contributions from community psychology. Health Promot Int 9:199–210

Hitchcock G, Hughes D (1995) Research and the teacher: a qualitative introduction to school-based research, 2nd edn. Routledge, London

Israel BA, Schulz A, Parker E, Becker A, Allen III A, Guzman JR (2003) Critical issues in developing and following community-based participatory research principles. In: Minkler, M, Wallerstein, N (eds) Community based participatory research for health. San Francisco: Jossey-Bass, pp 53–76

Israel BA, Schulz AJ, Parker EA, Becker AB (1998) Review of community-based research: assessing partnership approaches to improve public health. Annu Rev Public Health 19: 173–202

Jewkes R, Murcott A (1996) Meanings of community. Soc Sci Med 43:555–563

Kothari U (2001) Power, knowledge and social control in participatory development. In: Cooke, B and Kothari, U (eds) Participation: the new tyranny?. Zed Books, London, pp 139–152

Labonte R (1989) Community empowerment: the need for political analysis. Can J Public Health 80:87–88

Labonte R (1996) Community development in the public health sector: the possibilities of an empowering relationship between state and civil society, unpublished PhD dissertation, York University, Toronto.

Laverack G, Labonte R (2000) A planning framework for community empowerment goals within health promotion. Health Policy Plan 15:255–262

Leung MW, Yen IH, Minkler M (2004) Community based participatory research: a promising approach for increasing epidemiology's relevance in the 21st century. Int J Epidemiol 33: 499–506

Lewin K (1946) Action research and minority problems. J Soc Issu 2:34–46

Maguire P (1987) Doing participatory research: a feminist approach. School of Education, University of Massachusetts , Anherst, MA

Maguire P (1996) Considering more feminist participatory research: what's congruency got to do with it?. Qual Inq 2:106–108

McKnight JL (1987) Regenerating community. Soc Policy 17(3):54–58

McTaggart R (1991) Principles for participatory action research. Adult Educ Q 41:168–187

McTaggart R (1997) Participatory action research: international contexts and consequences. State University of New York Press, Albany

Meyer J (1993) New paradigm research in practice: the trials and tribulations of action research. J Adv Nurs 18:1066–1072

Minkler, M and Wallerstein, N (eds) (2003) Community based participatory research for health. Jossey-Bass, San Francisco, CA

Mohan G (2001) Beyond participation: strategy for deeper empowerment. In: Cooke, B and Kothari, U (eds) Participation: the new tyranny?. Zed Books, London, pp 16–35

Nilsen O (1996) Community health promotion: concepts and lessons from contemporary sociology. Health Policy 36:167–183

Oakley P et al (1991) Projects with people: the practice of participation in rural development. International Labour Office, Geneva

Parfitt T (2004) The ambiguity of participation: a qualified defence of participatory development. Third World Q 25:537–556

Park P (1993) What is participatory research? A theoretical and methodological perspective. In: Park P, Brydon-Miller M, Hall B, Jackson T (eds) Voices of change: participatory research in the United States and Canada. Bergin and Garvey, West Port, CT, pp 1–20

Park P, Brydon-Miller M, Hall B, and Jackson T (eds) (1993) Voices of change: participatory research in the United States and Canada. Bergin and Garvey, West Port, CT

Parkes M, Panelli R (2001) Integrating catchment ecosystems and community health: The value of participatory action research. Ecosyst Health 7:85–106

Peters M, Robinson V (1984) The origins and status of action research. J Appl Behav Sci 20: 113–124

Rappaport J (1985) The power of empowerment language. Soc Policy 16:15–21

Reason P (1994) Three approaches to participatory inquiry. In: Denzin, NK and Lincoln, YS (eds) Handbook of qualitative research. Sage, Thousand Oaks, CA, pp 324–339

Rifkin S (1996) Paradigms lost: toward a new understanding of community participation in health programmes. Acta Tropica 61:79–92

Robertson R, Minkler M (1994) New health promotion movement: a critical examination. Health Educ Q 21:295–312

Schwab M, Syme SL (1997) On paradigms, community participation, and the future of public health. Am J Public Health 87:2049–2051

Scriven A, Stiddard L (2003) Empowering schools: translating principles into practice. Health Educ 103:110–119

Tandon R (1996) The historical roots and contemporary tendencies in participatory research:implications for health care. In: de Koning K, Martin M (eds) Participatory research in health: issues and experiences. Zed Books, New Jersey, NJ, pp 19–26

Wallerstein N (1992) Powerlessness, empowerment, and health: implications for health promotion programs. Am J Health Promot 6(3):197–205

Wallerstein N (1999) Power between evaluator and community: research relationships within New Mexico's healthier communities. Soc Sci Med 49:39–53

Wallerstein N (2002) Empowerment to reduce health disparities. Scand J Public Health 30(Suppl 59):72–77

Wallerstein N, Bernstein E (1988) Empowerment education: Freire's ideas adapted to health education. Health Educ Q 15:379–394

Wallerstein N, Bernstein E (1994) Introduction to community empowerment, participatory education, and health. Health Educ Q 21(2):141–148

Williams G (2004) Evaluating participatory development: tyranny, power, and (re)politicisation. Third World Q 25:557–578

Whyte WF (1991) Participatory action research. Sage, Newbury Park, CA

Yeich S, Levine R (1992) Participatory research's contribution to a conceptualization of empowerment. J Appl Soc Psychol 22:1894–1908

Zakus JD, Lysack LC (1998) Revisiting community participation. Health Policy Plann 13(1):1–12

Chapter 5
The Process of Conducting an Ecohealth Research Project: A Participatory Action Research Approach

Contents

5.1 Introduction

The previous chapter discussed the theoretical and historical basis of community-based participatory research and some basic principles of participatory action research (PAR). It also cautioned against the uncritical adoption and use of concepts such as "community", "participation" and "empowerment". This chapter moves

C.Y. Dakubo, *Ecosystems and Human Health*, DOI 10.1007/978-1-4419-0206-1_5,
© Springer Science+Business Media, LLC 2011

beyond the epistemology and ontological basis of PAR, to discussing how participatory action research might be used to conduct an ecohealth research project in the field. In particular the chapter pays attention to the research process and takes ecohealth researchers and practitioners through a step-by-step process in the field of one case study. Some of the issues to be addressed include:

- How to gain entry into a community or research site
- How to form a transdisciplinary research team
- How to develop a research agenda, including choosing and developing data gathering procedures and tools with community members
- How to collect and analyse data, including using focus group discussions, follow-up individual interviews, and site visits; and
- How to put the research findings to action

Although the description of this research process is based on an ecohealth research project conducted in a community in Ghana, West Africa, the procedures are broadly described so they can be applied to other communities, especially those in developing countries, as well as in many small towns in the Western world, and in Indigenous communities. In describing the research process, the chapter draws on the author's experience in conducting an ecohealth research, and so there are times when the expressions in the text will be written from the first person perspective, especially in this chapter and the next two chapters, so as to illustrate the practicality of working in the field. A sample research schedule is presented in Table 5.1.

Table 5.1 Sample activity schedule for an ecohealth research project

Week #	Activity
1	Entering the community - Researcher meets with a gatekeeper or first contact
2	Gatekeeper introduces researcher to the community and elders, and other key stakeholders
2–3	Researcher visits potential members of a transdisciplinary research team
4	Researcher forms a transdisciplinary research team
5	Researcher organizes a training workshop for research team
5	Transdisciplinary team develops a research agenda
6	Transdisciplinary team visits community and meets with participants
6	Research team and participants set agenda for research activities
7–10	Research team commences focus group discussions and interviews
11–12	Research team conducts field visits and in-depth follow up interviews
13	Research team and participants collectively interpret and analyze data
14–16	Action planning and implementation of study results
17+	Monitoring and evaluation

5.2   Gaining Entry into the Research Community

In many cases researchers set of with a good idea of the site for their research. In the case of an ecohealth research project, a site may be chosen because of pre-existing information about an environmental problem (e.g. water pollution, land degradation) that adversely impacts human health, or a health problem (e.g. malaria, dengue, schistosomiasis) that is suspected to be influenced by environmental conditions of some sort, or a community just seeking to plan towards being a healthy community. Regardless of the scenario, many of these projects are located in communities that are guided by rules and norms. Hence a researcher, especially those coming from outside of the community, must find ways of seamlessly gaining entry into the community and given permission to conduct the research. The process of gaining entry into the community, explaining the purpose of the research and how the research will benefit community members can be a tricky exercise, especially for external researchers. Many communities in developing countries and in Indigenous communities have become weary of all the "good-will" studies that are being conducted in their communities, and the time and energy often required to respond to researcher interviews. Many of these communities have become weary because they see little change in their pre-existing conditions after the study is completed. Instead, what mostly happen is that the findings of the study make their way into prestigious journals, with no apparent benefit to the community. Participatory action research tries to bridge this gap by incorporating mechanisms for capacity building, education, and action. However, there is no guarantee that such outcomes will be achieved given the long timeframe often required to implement and monitor actions emanating from such research. In most instances researchers tend to work with tight budgets and short-time frames making it difficult to work with communities until learning consolidates.

Many communities in developing countries, especially Africa, have community leaders and gatekeepers who are usually suggested as the first point of contact when attempting to gain entry into a community. Sometimes it may be difficult having access to the gatekeeper or community leader. In such circumstances, it is better to contact a local government official who works in the community (e.g. an agricultural extension officer, a nurse, or teacher) to formally introduce an external researcher to the community gatekeeper. This first meeting is an important one and provides an opportunity for researchers to explain their mission, the purpose of the study, how it will benefit the community and a request to meet with the chief or figurehead of the community "with" and elders of the community as well as any key stakeholders.

The meeting with the chief and elders is equally very important as this is when the researcher has to clearly articulate why that site was chosen, how the research will involve community members and most importantly how the research will change, address, or respond to the conditions currently facing the community. The researcher also uses this first meeting to examine how their objectives align with that of the community. Does the community consider the environment or health issue to be investigated a problem? As discussed in the previous chapter, an ideal PAR project requires that the researcher and the community mutually identify and agree on the research problem to be investigated, but because many communities lack this initial drive or resources to initiate a study, it is mostly the external researcher who approaches the community and initiates the discussion, hoping to build some common interest. It is also important for researchers to state these objectives as tentative as these will probably need to be refined with community members and the research team when the research commences.

Following this discussion, the researcher formally requests permission from the chief, elders and key stakeholders to conduct the research in the community. This request also includes permission to meet with members of the community, request for their cooperation and participation in the study, permission to use community facilities and resources, and permission to take pictures, videos, or samples of any research material outside the community. In response, the chief consults with his elders and grants or refuses permission, or request for further clarification of certain issues, including more information on why their community was chosen, who will be responsible for the costs of the implementation of any actions coming out of the project, how will project outcome influence public policy, how will the project be sustainable over the long term, and clarifications on issues of monitoring and evaluation. After satisfactorily responding to these questions the chief may then grant permission for the research to proceed.

Researchers who follow these procedures to gaining entry into study communities may find a smooth progression in their research, especially in building trust with study participants and also building momentum and enthusiasm in the study. Some external researchers do not follow this route because they are either unaware of it, find it unnecessary, or think it might suffice with government approval or an affiliation with an academic institution in the region or country. The failure to follow the necessary procedures or precautions to gaining entry into a study site may result in lack of cooperation from community members, difficulty building trust, or forcing the researcher to terminate the research project prematurely. However, in some circumstances, entry into a community may be facilitated when there is already an on-going research project or international development projects in the community, with which the researcher is affiliated. In such circumstances, the process of gaining entry is less tedious.

5.3 Forming a Transdisciplinary Research Team

One important pillar of the ecohealth approach is transdisciplinarity. An ecohealth research project brings together researchers from various disciplines, including health, environment, anthropology, sociology, toxicology, among others. This group of researchers collaborates with representatives from the community and other key stakeholders to form a transdisciplinary research team. Together the research team develops a research agenda, delineates the problem to be investigated, determine data gathering procedures, identify participants, and collect and analyse data. In western countries, it is relatively easy to find professionals from relevant disciplines to come together to form a transdisciplinary research team. However, in a developing country context, where such expertise is limited, forming a transdisciplinary team with the required representation can be a difficult task and requires a bit of creativity.

In such circumstances, a transdisciplinary team could be formed by bringing together local sector experts such as foresters, mining engineers, fisheries officers, and local government officials who work in the community, including agricultural extension officers, community health nurses, teachers, and social workers. These sector experts and local government officials work closely with communities and tend to have good understanding of community problems and sociopolitical dynamics in the community. In the West African case study, my initial contact was with the co-ordinator of an international development project, a government official with the department of education. The coordinator has a long working relationship with the community and so was a good gatekeeper in assisting me gain entry into the community. Given that he is also a government official, he was instrumental in identifying key government departments and sectors that worked in the community as potential members of the research team. These departments included the ministry of health, ministry of food and agriculture, the department of forestry, the institute of adult education, and representatives from other international development projects. Representatives from these institutions, together with the principal investigator (i.e. myself), and two representatives from the community – the head teacher for the village Primary school, who is also the community leader, and the headmistress for the village Junior Secondary School, constituted the core of the transdisciplinary research team.

Within the context of participatory action research, all members of the research team are colleagues, partners and co-researchers, who are suppose to be actively involved in all stages of the research process. However, in a developing country context, given that not all members of the research team will have the same level of research skills as the external principal investigator, it is encouraged that the lead investigator organize a mini-training session or workshop to share some data gathering tools, resources and processes with the team. For example, in the Ghanaian case study, a series of workshops and training sessions were organized to share data gathering processes such as focus group discussions and strategic planning processes with members of the research team. These preliminary training sessions' ensure that members of the research team are on the same page and are well-acquainted with any specialized data collection procedures that may be used. Such workshops also provide members of the research team with useful research skills, as well as create a space for the team to begin to work together and learn about how their respective mandates and duties complement one another in the study community. Inter-departmental collaboration is not a common practice in many settings, hence extra effort is usually required to ensure a well-functioning transdisciplinary research team.

These preliminary workshops also serve as good venues to draft a research agenda. The research agenda outlines the roadmap for the research. It identifies the issues to be investigated, the goals and objective of the research, how participants will be chosen and sets a timeframe for the study. This draft agenda will be presented to community members participating in the study for their input and feedback. It will also outline a schedule of activities, including times for focus group discussions, field trips, and availability of participants before being finalized. This collective development of the research agenda encourages ownership and active engagement in the research process. It allows community members and the research team to perceive the research as "our" research and not "their" research.

In most western countries, while workshops with the research team may not be about training and acquainting members with research methods and data gathering procedures, it is still important that the research team gets together to familiarize themselves with each others disciplinary backgrounds, explore the synergies and common interests, and develop a research agenda prior to commencing the study.

5.4 Recruiting Research Participants

One primary objective of participatory research is to ensure that all voices are heard during the research process. Participatory researchers seek active representation of all facets of society, particularly, all stakeholders who one way or another might be affected by the issue under investigation. This includes the vulnerable and the marginalized. However, often, as outside researchers going into a community, it is always difficult to identify all these relevant voices in a timely fashion, let alone actively involve them in all stages of the research process. Similarly, those who may be especially affected by the problem under investigation are always too busy trying to secure their livelihood (Rifkin 1994). Also, some researchers find it easy to recruit participants from existing organized groups in the community, such as Christian mothers organization, Muslim women's groups, tree planters association, youth groups, among others. The concern is that such groups are usually organized around certain core goals, values, and ideologies (political, religion, etc.) and are not necessarily representative of the community. The question then is, how should researchers go about choosing a representative group of community participants in an environmental health research project that has community-wide implications?

During my fieldwork, the research team indicated the need to work with two separate groups of participants: a men's group and a women's group. They thought this was important within the cultural context, as women will not speak candidly about their health concerns in the presence of men, for fear of being perceived as the "lazy wife". Similarly, men would not be comfortable discussing health concerns before women as this may be a sign of weakness. Such rationalization led the team to form separate groups for men and women and to collect gender-disaggregated data. While it is alright to form groups based on gender, care must be taken not to essentialize men and women's experiences, that is, interpret these as belonging to specific genders, but instead take steps to illustrate how these experiences are shaped by their multitude identities, roles, and tasks. With respect to gender-based environmental health research, Kettel (1996) points out that, the biophysical environment affects the health, social and economic lives of men and women differently. As men and women go about their daily activities, they acquire different familiarization with their social and physical environments, providing them with a view that is characteristic of the roles, and spaces they occupy. These distinctive activities and knowledges must be clearly acknowledged in the research analysis and evaluated in context with other changing identities and situations.

In this study, participants for the men and women's groups were selected using a variety of strategies including a snowball sampling strategy, with primary contacts emanating from the two community representatives on the research team, who happen to be a male and a female. In addition there were public announcements of the research in the community and interested participants were asked to contact members of the research team. A participant information sheet allowed the research team to keep track of who who was interested, and why. These information sheets allowed for a good sample to be selected for both men and women's groups. Although this was the formal group for the study, other community members dropped by occasionally to listen to the discussions. It is also important to realize that this is the "public" group that is willing to provide public accounts (Scott 1990) of their understanding of the problem under investigation. Beyond this are "private" accounts from other relevant stakeholders which must be tracked down through other data gathering procedures such as in-depth interviews and follow-up interviews.

The research team met separately with the men and women's groups to explain the purpose of the study, seek their consent to participate, explain the duration of the study and the time commitment required. Participants also used this first meeting to review the research agenda and identify the days, venues and times for focus group discussions and field visits. This initial meeting was also used to gather basic demographic data from participants including age, marital status, level of education, religion, occupation, number of children and approximate family income.

Given that many researchers gain entry into the community through local leaders and gatekeepers who play a key role in identifying study participants, the researcher, to some extent, has very little control over who participates. For example, during my fieldwork, while I was interested in reaching all relevant stakeholders, including the marginalized, they probably could not be involved because of competing priorities, did not belong to any community organization, did not have clean clothes to join group discussions, or were too afraid to speak publicly. As Cornwall and Jewkes (1995: 1673) point out, 'unless a definite political commitment to working with the powerless is part of the [research] process, those who are relatively inaccessible, unorganized and fragmented can easily be left out'. The challenge this situation presents is that by working through local power structures, there is the tendency for the research to be manipulated towards the agendas of local authority. On the other hand, working against local authority could weaken the potential impact of the research outcome and increase alienation, inequities and marginalization after the research is completed (Cornwall and Jewkes 1995).

5.5 Data Gathering Processes

Like many transdisciplinary research endeavours, a well-crafted ecohealth research project can generate a wide spectrum of information, including quantitative and qualitative data, narratives, scientific and lay perspectives. As such, it is important to identify effective data gathering processes that will capture all the relevant data

being produced. In addition, participatory action research emphasizes the importance of both the research process and outcome. Hence data gathering should not
be extractive, but should also constitute a learning environment, where people can
dialogue and develop new insights about the problems facing their communities and
also build community capacity.

There are a number of data gathering tools that can be used for ecohealth
research. Usually a combination of individual and group methods is recommended,
since not all issues can be addressed effectively through group processes. Individual
methods include structured and unstructured interviews, in-depth and follow-up
interviews, and site visits. Common group processes are workshops and focus group
discussions. Recently, some innovative group processes such as the search conference and strategic planning techniques have been used to help communities plan
for a healthy community (Dakubo 2004, 2006). The strategic planning process is
commonly used in community economic development and corporate planning circles to help communities and business enterprises move towards achieving a set of
economic development or business goals. Its use in ecohealth research projects is
quite limited. Nevertheless, community strategic planning processes enable communities to articulate their vision for a healthy lifestyle and to evaluate what major
environment and health problems are blocking the realization of that vision. They
then identify and implement the necessary steps to making the vision of a healthy
community a reality. Technical and secondary data also need to be collected from
all relevant departments and institutions. Some of these data gathering processes are
discussed next. Figure 5.1 shows a research process that was used in an ecohealth
research project (Dakubo 2004).

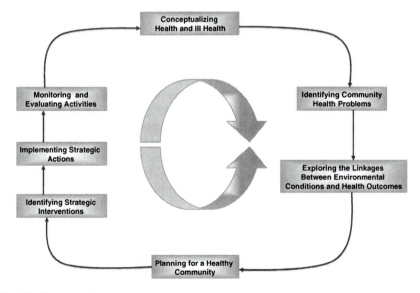

Fig. 5.1 The research process

5.6 Focus Group Discussions

Focus groups discussions are qualitative research processes that bring a group of participants usually (between 6 and 12 people) together to gather their views, opinions and ideas about a particular topic. These discussions are facilitated by the researcher or the research team and usually are interactive in nature, allowing participants to discuss their views about a particular topic in a natural and non-threatening environment. Compared to one-to-one interviews, focus group discussions are able to gather in-depth information on a variety of issues within a relatively short timeframe. Their interactive nature allows participants to build on one another's ideas and generate novel ideas that would not have emerged in an individual interview. Participants also challenge each others views and force the group to be thorough and logical in their thinking, thus increasing the validity of such discussions.

Within the context of ecohealth research, focus group discussions could be used to understand how people conceptualize health, what they consider to be indicators of good or poor health, what they perceive to be the main causes of poor health, and how they associate environmental conditions with health problems in the community. Data from such focus group discussions can then be contrasted with "expert" constructions to determine the congruence and differences between these concepts, and the implications for health intervention in the community. As discussed previously, not all issues are amenable to focus group discussions, as such it is important for the researcher or research team to be particularly vigilant for issues that may be alluded to that need further investigation. These could be explored through follow-up in-depth interviews so as to to avoid wrongful interpretation of group discussions.

5.7 Follow-up and In-depth Individual Interviews

Follow-up and in-depth interviews with individuals are useful in gaining further insight into issues that are raised during group sessions but cannot be discussed in a broader group session. Focus group discussions may identify certain key stakeholders who are not part of the study participants, but whose views would be relevant in responding to the problem at hand. These individuals are then contacted and requested to grant an in-depth individual interview. Alternatively, during a focus group discussion, the facilitator may observe that an individual is holding on to some information that may be relevant to the study. In such circumstances, a follow-up interview is requested to gain a better insight into the discussions.

In addition, some traditional knowledge systems are held sacred, some traditional healers or medicine men abhor group sessions and would not participate in focus group discussions, although they would be willing to grant personal interviews. Individual, in-depth interviews with traditional healers could reveal a wealth of knowledge about the various medicinal plants in the community and some of the conservation methods used to preserve them. Such interviews can also reveal the extent to which these plants have become extinct based on historical narratives.

Follow-up and in-depth interviews are not limited only to community members, but can also be used with members of the research team. As discussed previously, some sector experts and government officials are not accustomed to working together in transdisciplinary settings. As such, members of the research team may have observations that could be useful to the research, but because of poor team relationships or power dynamics within the research team, such information may fail to become available through group processes. Hence, follow-up and in-depth interviews are ideal for gathering these views to complement, other sector or departmental insights that might have been gathered through other means.

5.8 Strategic Planning for a Healthy Community

Strategic planning is a process that allows communities and organizations to define their vision and identify the necessary steps, actions and resources needed to achieve that vision. The process brings people together to establish common goals and search for a desirable future. Strategic planning process (SPP) is designed to tap the unique abilities, strengths, and knowledge base of participants in a way that builds on group dynamics and group learning. It allows participants to look beyond their immediate problems and obstacles and to recast their efforts to a desired future (Spencer 1989). Strategic planning is widely used by businesses, corporations, and communities. Except for community health planning, its use within the spheres of ecohealth research is very limited.

Although there are variants to the strategic planning process, they mostly proceed through a series of up to five cyclical steps (Spencer 1989). The first step involves mapping out a community's vision and aspirations for a desired future (e.g. a healthy community); the second identifies the underlying obstacles preventing the realization of that desired vision; the third step identifies broad strategic directions that can be implemented to overcome the identified obstacles; the fourth step follows through with the identification of specific systematic actions to implement the strategic directions outlined in the previous step; and the last step outlines an implementation timeline or schedule for the activities and follows through with implementation. These steps are not linear, as previous steps can always be revisited as the planning proceeds and new insights are gathered. In the sections below, I describe a strategic planning process that was adapted from Spencer (1989) and used to help a rural community in Ghana plan for a healthy community (Dakubo 2004, 2006). To help facilitate the discussion, each step of the strategic planning process is preceded by a focus question, which was designed by the research team. Community strategic planning was conducted separately for the men and women's groups.

5.9 Steps to Planning for a Healthy Community

5.9.1 Step 1: Mapping Out a Vision for a Healthy Community

The first step of the planning process maps out a desired future for the community. Participants describe the features and conditions they would like to experience in their vision of a healthy community. Participants are encouraged to identify practical, concrete, and attainable vision elements. A realistic vision challenges participants to strive towards making this vision a reality. In the context of the ecohealth study, the research team posed the focus question below and facilitated the group discussion:

> Imagine we are five years into the future, and all the people living in this community are feeling well and healthy, with no more environmental problems or no health problems. Describe the noticeable features and conditions you would want to see in this healthy community?

In responding to this question, participants will identify and describe what they perceive to be the attributes or qualities of a healthy community. The research team facilitates the discussion by probing for elaborations on certain responses. Participants will then categorized their responses into themes, discuss, and prioritize or rank them.

5.9.2 Step 2: Analysing Underlying Obstacles and Barriers to Achieving a Healthy Community

During the second step of the planning cycle, participants are guided through a focus group discussion to identify all possible factors they see as obstructing or could obstruct the realization of their vision. These obstacles are usually wide-ranging, span beyond the health and environment sectors. Some of these obstacles could relate to local power struggles, uneven access to and use of natural resources and health services, constraining policies, and other socio-political and economic constraints. To be effective, obstacles must be addressed from the root cause, and so participants should be guided to examine the underlying causes of these obstacles. A sample focus question could be:

> What obstacles or roadblocks are preventing us or could prevent us from achieving our vision of a healthy community?

5.9.3 Step 3: Identifying Appropriate Strategic Directions for a Healthy Community

This step of the strategic planning process identifies creative and innovative strategies and programs designed to overcome the obstacles and barriers identified in the previous step. In identifying strategies, proposals and programs, participants must ensure that these are feasible and can be implemented with minimal external help. Participants are guided to respond to the following focus question:

> What activities must we undertake to overcome the barriers and roadblocks identified in our previous discussion?

5.9.4 Step 4: Identifying Systematic Actions and Assessing Community Capacity

This step of the strategic planning process calls for the identification of practical and attainable actions that can be implemented to fulfil the broad proposals and programs outlined in the previous step. It also requires that community members assess their strengths, weaknesses and the resources available to them to be able to implement the actions recommended. The following focus question guides this discussion:

> What practical and attainable actions can we begin to implement so as to achieve the proposals and programs identified previously? What are our strengths, weaknesses, opportunities and threats (SWOT)? Who are our allies and partners in this effort, both internally and externally?

This question allows participants to identify feasible actions to be implemented so as to achieve their vision of a healthy community. It also allows participants to engage in a SWOT analysis, and a frank and honest discussion about the resources and assets available to them to undertake these activities. The objective here is to ensure that actions are sustainable and will not create dependencies. Participants can draw on external allies to support them with the implementation of proposed activities. While the focus is on progress, it is also important to take stock of the threats and weaknesses that could prevent participants from implementing the proposed actions. Compared to the earlier barriers discussion, these threats and weaknesses are focused and relate to the proposed activities to be implemented.

5.9.5 Step 5: Developing an Implementation Schedule and Carrying Out Proposed Actions

The strategic planning process concludes with the development of a timeline for the implementation of all activities identified in the previous step. The implementation timeline outlines activities such as: what will be done, by who, where, when, what resources are needed, when will the activity be completed, who will monitor it, and how? At this stage, participants begin to feel a sense of pride, especially in developing country settings where people perceive themselves as helpless victims whose progress is continuously dependent on external expertise and assistance. The ability to develop and implement a plan that depicts a desired future for the community is a healthy feeling by itself.

5.10 Site Visits

Site visits are particularly useful when conducting an ecohealth research. They provide an opportunity to validate issues that have been discussed during focus group discussions. For example, during the West African study, I journeyed with women and men to their farms. These farms are usually located in the outskirts of the community and can take up to 3 hours to reach there. The walk to the farm is an educational trip, in itself, as farmers take this opportunity to explain the ecological significance of various plants and their medicinal uses. I also visited the sites of small business enterprises such as, basket weaving, charcoal making, potteries, and carving shops. These visits did not just illustrate how locals interacted with the biophysical environment, but also demonstrated the differentiated environmental knowledge systems between men and women, and between the elderly and the youth. The visits to the work environments allowed me to witness the varying occupational health risks of these small business enterprises that community members engage in, and the protective and coping strategies they employ.

5.11 Secondary Data Sources

In order to complement data and information from group discussions, site visits, and individual interviews, it is important to collect secondary data from other sources that are relevant to the problem under investigation. For example, relevant data from the Department of Health department may include: disease incidence and distribution patterns, and morbidity and mortality data. Environment and climate-related data may be obtained from the Departments of Environment, while food production trends could be obtained from the Ministry of Food and Agriculture. Other departments or ministries could also provide information related to education, and economic trends and forecasts. Secondary data are useful during data analyses, and help put quantification, distributions, and patterns to some of the qualitative data gathered through group discussions and other data gathering procedures.

5.12 Data Management and Analysis

As discussed previously, ecohealth research tends to produce large quantities of both qualitative and quantitative data, making it even more important to find effective ways of collecting, organizing and managing this data. The careful collection and organization of data makes data analysis less stressful and relatively straightforward.

To keep track of activities and information in the field, a number of tools may be used, including journal entries, photographs, audiotape recordings of group discussions, videotape recordings of group discussions and farm visits, and diagrams of activities at site visits. To ensure a common understanding of events and the data collected, it is a good practice to recap all the discussions and activities at the end of each day, summarize the main points, prioritize them, and identify those that need further clarification. Prior to the start of the next day's activities or sessions, it is helpful to remind participants where they left of in the previous day so as to ensure continuity.

The interpretation and analysis of data in participatory research projects is supposed to be a collaborative exercise between the research team and the study participants. In order not to complicate matters, it is helpful to conduct data analysis of all data gathered at the end of the day, week, and month. In analyzing data, a number of methods can be used, including recapping key issues of each discussion, identifying the broad areas of consensus or difference, and identifying emerging themes and constructs around which questions were asked. Data analysis should be a cyclical and dynamic process, with insights from previous stages of the research used to inform subsequent stages of analysis. Collective data analysis creates an environment for learning whereby participants learn to make associations between phenomena; for example being able to make connections between water collected in a pond, how that creates a favorable breeding ground for mosquitoes, and how this, in turn, contributes to the incidence of malaria. Such insights then allow participants to begin to take the necessary steps to protect their health, while undertaking appropriate land use practices. However, while data analysis may be a joint effort between the research team and study participants, the write-up of the study is the responsibility of the lead investigator, and there are chances that not all information will be interpreted as accurately as was discussed in the field. Hence, it is always a good idea to run the report by the community for cross-checking or state a disclaimant.

5.13 Conclusion

This chapter outlines the process of conducting an ecohealth research project in the field. Research is a messy exercise and care must always be taken to ensure that all caveats are covered. With ecohealth research, in particular, the requirement to form a transdisciplinary research team is a difficult undertaking, especially in developing country contexts, or with student researchers, who usually have limited networks with other potential collaborating institutions. In addition, gaining entry into a study site, building trust, and selecting a representative sample of community members to work with can be fraught with a number of challenges and require tactfulness. Finally, the choice of data gathering procedures has to be carefully considered so as to provide opportunities for participants to learn from the process, acquire some research skills, and be better positioned to investigate future community health and environment concerns. The chapter also discusses innovative strategic planning processes that were used to help a rural community in Ghana plan towards achieving their vision of a healthy community. In the next two chapters, we will examine the findings of this research project, and how this could serve as a model for other communities.

References

Cornwall A, Jewkes R (1995) What is participatory research?. Soc Sci Med 41:1667–1676

Dakubo C (2004) Ecosystem approach to community health planning in Ghana. EcoHealth 1: 50–59

Dakubo C (2006) Applying an ecosystem approach to community health research in rural Northern Ghana. Unpublished PhD dissertation, Department of Geography and Environmental Studies. Carleton University, Ottawa, Canada

Kettel B (1996) Women, health and the environment. Social Sci Med 42:1367–1379

Rifkin S (1994). Participtory research and health. In: Proceedings of the international symposium on participatory research in health promotion, Liverpool School of Hygiene and Tropical Medicine. Liverpool, UK, Sept. 1993

Scott JC (1990) Domination and the arts of resistance. Yale University Press, New Haven, CT

Spencer L (1989) Winning through participation. Kendall/Hunt Publishing Company, Dubuque, IA

Part III
Case Studies: Application of the Ecohealth Approach

Chapter 6
Applying an Ecosystem Approach to Community Health Research in Ghana: A Case Study

Contents

6.1 Introduction

This chapter discusses an ecohealth project that was conducted in a small rural community in the Upper West Region of Ghana. Ghana is located along the West Coast of Africa and has often been referred to as an "island of peace" because of the long-standing tranquility that exist in the country, compared to others in the west africa sub-region. Ghana has population of about 23 million, with about half the population living in rural areas. The population growth rate is estimated at 2.6% with a total fertility rate of 4.0 (WHO Country Profile).[1] Administratively, the country is divided into ten regions and 170 District Assemblies. Each District Assembly is responsible for developing, planning and mobilizing resources to implement programs and proposals for the development of its region. Besides being divided into administrative regions, the country is well-known for the stark difference between the Northern

---

[1] World health Organization Country profile. http://www.who.int/countries/gha/gha/en/ Accessed May, 10th 2010.

C.Y. Dakubo, *Ecosystems and Human Health*, DOI 10.1007/978-1-4419-0206-1_6,   89
© Springer Science+Business Media, LLC 2011

and Southern parts of the country in terms of human development, infrastructure development, and health services. This vast difference sometimes makes it more meaningful to characterize the country by its geographic divisions of North and South than by its administrative regions, especially when examining and responding to issues of unequal development. A recent World bank report observed that, while there seems to be a general trend of improvement in the health outcomes of many Ghanaians, this trend is masked by vast health disparities between the north and the south, between rural and urban regions, and between various socioeconomic groups (World Bank 2003). The study suggests that the largest differential in health outcomes is by region of residence, with the three regions in the North: the Upper West, Upper East, and Northern regions consistently faring poorly in many health outcomes, compared to their Southern counterparts. For example, the under-five mortality rate in these three northern regions is 2.5 times higher than in the Greater Accra Region, which is capital region. Also, whereas the mortality rate in the Greater Accra Region is 62 deaths per 1000 live births, it is 171 in the Northern Region, 156 in the Upper West Region and 155 in the Upper East Region (World Bank 2003: 8). Also, the Greater Accra Region which holds about 12% of the national population accounts for 42% of total public doctors and 18% of hospital beds in the country. In contrast the northern regions with 20% of the population account for 6% of total doctors and 14% of the total hospital beds (ibid). Similarly, education, literacy, and income levels are lowest in rural areas and in most northern parts of the country.

The existing socioeconomic and health disparities and inequalities characterizing Northern and Southern Ghana can be traced back to the colonial era, when many resources were proximally located near mineral deposits in coastal southern regions to the detriment of the rest of the country. Railroads, factories, hospitals, universities, and major infrastructure were all based in the South, with the North serving as a labour pool (Songsore 1983). Since gaining independence in 1957, there have been increasing attempts by different governments to bridge this development gap, although this has proceeded slowly.

From a political and economic standpoint, Ghana has made steady progress since 2000. Politically, Ghana has made smooth transitions between governments since 2000, while economically, the country's gross domestic product (GDP) has steadily increased from 3.7% in 2000 to 7.3% in 2008 (Government of Ghana 2008). The Government's development agenda is to transform Ghana into a middle-income country with a GDP per capita of at least 1000 USD by 2015. The strategies for achieving this include investing in human capital, strengthening private sector growth, and providing good governance (WHO Country Cooperation Strategy)[2]. In addition, the country has made steady progress towards achieving some of the Millennium Development Goals (MDGs), especially those related to education and poverty reduction, although more progress is needed in health-related MDGs. For

---

[2]WHO Country Cooperation Strategy http://www.who.int/countryfocus/cooperation_strategy/ccsbrief_gha_en.pdf. Accessed May 10th 2010.

example, while the trend for maternal mortality ratio seems to be declining, there are still 451 deaths per 100,000 live births, and 80 out of 1000 children dying before their fifth birthday (2008 Demographic Health Survey).

Currently, Ghana is going through an epidemiological transition. For several decades, Ghana's disease profile has remained unchanged, with communicable and infectious diseases, undernutrition and poor reproductive health topping the list of hospital attendances. The major causes of child mortality include malaria, diarrhoea, respiratory infection, and neonatal conditions. In particular, malaria has been identified as a primary cause of poverty and low productivity in the country. Malaria alone accounts for over 44% of reported outpatient visits and an estimated 22% of under-5 mortality in the country (WHO Country Report for Ghana). However, with changing lifestyles in urban areas, non-communicable diseases such as hypertension, diabetes, cancer and mental illnesses are on the rise, as well as increase in tobacco use, alcohol consumption and other substance abuse. Compared to other countries in the region, Ghana still has a low incidence rate of HIV/AIDS. In an attempt to improve health conditions in the country, the health sector, through its Programme of Work (2007–2011), closely links health improvement efforts with poverty reduction through the Growth and Poverty Reduction Strategy (GPRS II). To complement this strategy, the adoption of an ecohealth approach to health sector programming is worthwhile, given that infectious diseases continue to be the leading cause of many health problems in the tropics. Although environmental conditions play an important role in many health problems facing African countries, the linkages between health and environment do not feature prominently in public health promotion and environmental management policies. It is against this backdrop that an ecosystem approach to community health research was conducted in a Ghanaian community to assess the extent to which the ecohealth approach can contribute to responding to the health and environment challenges facing this community.

6.2 Health and Environment Challenges of the Study Community

Given the above description of the health disparities between Northern and Southern Ghana, and rural and urban regions of the country, the community in which this study was conducted embodies the two curses: a rural northern community located approximately 8 km northwest from Wa, the regional capital of the Upper West Region. The community has a population of about 4,500 people with up to 90% of the population engaged in subsistence agricultural. Although many inhabitants engage in agriculture, food production is still low, and low crop production has been attributed to factors such as depleting soil fertility, use of outmoded farming equipment, unfavourable weather conditions, poor marketing outlets and lack of credit to support farming activities. In addition to agriculture, community members engage in smallscale enterprises and income generating activities such as pot making, *pito*[3] brewery, carving, charcoal making, blacksmithing, and petty trading.

Like many rural communities in the region, the health status of community members is relatively poor, with the major diseases being malaria, acute respiratory infections, and diarrhoea. Compounding these health problems are poor sanitation, lack of clean water supply, low literacy levels, and high levels of poverty. The village has one health centre that provides primary care to residents, and refers major cases, especially those requiring surgery to the nearby regional hospital.

Compared to urban areas, rural areas tend to experience high levels of marginalization due to their peripheral location from the core centres of political action. As such, many rely on the natural environment as the main source of livelihood. As has been discussed previously, this environment is both a source of health and a source of disease, and this connection is well recognized by many local people who interact closely with their biophysical environments. What is probably not well understood is how external political and economic factors, and internal micro-politics and power dynamics shape community members interaction with the biophysical environment and consequently differentially impact their health and well-being. Also because of the continuous marginalization and perceived lack of capacity to change their living conditions, many rural communities tend to accept their poor health status and other challenges as given, without questioning why such poor conditions continue to prevail in their communities despite increased attempts to live in a healthy community.

As discussed in previous chapters, helping communities understand the underlying causes of poor health, and the factors that shape their relationship with the environment and produce various health outcomes go beyond the purview of the biomedical model and health sector. They also fall outside the domain of traditional applied research approaches that do not build in learning and social action as key components of the research process. In addition, these issues are not amenable to single-disciplinary investigations given that the health problems facing this community have historical antecedents and are shaped by social, political, and economic factors that serve to marginalize the community, and continuously place it in a cycle of poverty and poor health. Instead the health challenges facing this community seem to be very much aligned with the principles, goals and objectives of the ecosystem approach to human health where the use of a participatory action research, with a transdisciplinary team of researchers and local actors are able to explore the complex ways in which social, political, economic, and ecological factors interact to influence the health outcomes of the community. Through participatory action research, community members are able to become active agents in searching for the solutions to their environment and health problems.

---

[3]Pito is a local beer made from guinea corn.

6.3 Making Use of an Ecosystem Approach to Community Health Research

Based on the rationale outline above, this Ghanaian case study makes use of an ecosystem approach to community health research, which not only focuses on understanding the underlying factors responsible for poor health in the community, but also seeks to encourage the improvement of health through the adoption of sustainable ecosystem management practices. The ecosystem approach to health recognizes the importance of human agency in making conscious decisions and taking the necessary actions to improve individual health, but encourages the integration of such perspectives within reciprocal relationships between people and their social, economic and biophysical environments. In addition, being cognizant that many of the health problems facing this community are the result of political, economic and social marginalization, this study combines the ecosystem approach with a political ecology analytic framework to counter dominant, yet simplistic explanations about the persistence of poor health status in many rural Ghanaian communities. Many explanations for the persistence of poor health in rural communities, too often than not, continue to border on blaming the victim, and attributing poor health to people's reckless attitudes, poor hygienic practices, and ignorance of the rural folks. Some explanations blame poor health in rural African communities on lack of human and financial resources and limited international aid (Aidoo 1982). But as Aidoo points out such explanations reduce health concerns to resource-based problems, failing to question the reasons underlying the limited human and resource scarcity in the first place. Also, blaming rural people for engaging in poor health practices blames the victim and fails to situate the emergence and persistence of rural health problems within broader historical, political and socioeconomic contexts, and examining how colonial policies and economic adjustment reforms could mediate such poor health outcomes.

In addition to the above concerns, an ecosystem approach advocates a people-centred approach to health research and development. As such this study made use of a participatory action research strategy with the view to involving community members in all stages of the research process and also generating critical consciousness among participants about their health situation and how to contribute in finding effective solutions to them. Compared to traditional research approaches, community-based participatory action research is a suitable alternative that gives voice and power to those affected by a problem, to influence the research process and take action to respond to those problems (Maguire 1996; Wallerstein 1999).

This project therefore combines three theoretical frameworks: the ecosystem approach to health, a political ecology analytical framework, and community-based participatory action research. The integration of the three approaches provides a comprehensive overarching analytical framework from which to examine community health concerns from an ecosystem perspective, critically explain the spatial and social basis of the health disparities in the community, and at the same time generate dialogue and awareness about the factors influencing community health and how to respond.

6.4 Forming a Transdisciplinary Research Team and Setting the Research Agenda

A key feature of the ecohealth approach is to make use of a transdisciplinary team of researchers who them collaborate with all relevant stakeholders to develop a research agenda. Ideally a transdisciplinary research team must include professionals from the natural, social, and health sciences so as to be able to examine and interpret the research problem from an integrated perspective. However, in rural community settings, it can be difficult finding professionals from the various disciplines, requiring that concept of transdisciplinarity be adapted to ensure a true representation of all stakeholders working in the community whose activities influence community health and well-being one way or another.

In this study, the research team comprised of two community representatives and all relevant government departments working in the community, including forestry, agriculture, health, and education. I led the team as a principal investigator. The research team met to develop a research agenda following the procedures described in the previous chapter. Two groups of community participants were formed, a men's group and a women's group. The research team met with participants to finalize the research agenda and to outline a schedule for the research. Focus group discussions, workshops, in-depth interviews and field visits were used to collect data over a 6-week period. Separate group discussions were held for the men and women's groups. A strategic planning process, described in the previous chapter, was used to map out a vision for a "healthy community", and to assess community strengths, resources and constraints to achieving its vision. The findings of this planning session will be described in the next chapter.

Prior to planning for a healthy community, the research team thought it was appropriate to start from the basics and find out how community members conceptualized health and poor health. This was important because, the team was careful not to impose our definition of health on community members. Instead the team wanted community members to articulate these concepts from the perspective of their respective identities and social roles. In addition, it was important to understand what community members perceived as the major health problems facing the community and what they thought were the underlying factors and causes. Hence the objectives of the study were to explore local perceptions of health and ill health, understand the major health problems facing the community, explore the factors facilitating the occurrence of these health problems, and engage community members, government officials, and all relevant stakeholders in a joint participatory planning process to identify possible intervention strategies leading to the development of a healthy community.

6.5 Findings of the Study

6.5.1 Community Members' Perceptions of Health and Poor Health

Most often than not, as public health researchers, we tend to impose our definitions and conceptions of health and ill health on communities or research participants. These definitions reflect our western training, which makes predominant use of biomedical conceptions, and sometimes differ from local and Indigenous perspectives. In this study, the research team deemed it important to let community members express their own views of health, and what they considered to be poor health or ill health. These discussions were explored in separate focus group discussions with both men and women groups. Some of these responses are presented in the Table 6.1.

For the most part, participants' conceptions of health differed significantly from the biomedical view of health. Health was described as the ability to fulfill social responsibilities, access to basic services, meet personal needs, fulfill societal roles, and cope with everyday life circumstances. Participants conceived health from a variety of perspectives, including ecological, psychological, emotional, physical, spiritual, and access to social support networks. Most conceptualizations of health were contextualized and grounded in peoples' lived experiences and societal roles as husbands and wives, and mothers and fathers. Poor health, on the other hand, was conceived as the inability to fulfill one's duties in the family or in the community, constant worrying about how to meet one's daily needs, and feelings of unhappiness. Very few people referred to disease in their descriptions of poor health.

There were slight variations in how women and men conceived health and ill health. Their views were directly linked to their gender-related duties. For example, many male participants explained health in relation to their roles as household heads and husbands. They considered themselves healthy to the extent that they are able to provide food and shelter for their families, and meet the social obligations of a man in the community. Poor health, in their view, was the inability to carry out one's responsibilities as a male figurehead, both at home and in the community.

Women, on the other hand, conceived health in terms of their roles as caregivers, wives, and mothers. Health was seen as the ability to take good care of their children, and be seen by society as good and caring wives and mothers. Women expressed the additional pressure to stay healthy or silent about their health concerns in order not to disrupt the smooth functioning of the family, or for fear of not meeting the expectations of a "good wife." Women's expressions of poor health were mostly psychosocial: "worrying too much," "inability to sleep," and lack of social support networks.

These expressions of health point to the importance of broadening public health conceptualizations of health to reflect such social, cultural, and ecological dimensions, especially in developing countries and Indigenous communities contexts. While such notions of health are embodied in the World Health Organization's

Table 6.1 Men and Women's conceptions of health and poor health[4]

Conceptions of health	Quotation	Participant characteristic
Linked to social roles	*Health is being industrious and putting laziness aside. Healthy men pick up their hoes, go to the farm, and come home at dusk with food for their families. If a man is lazy and fails to provide for himself and his family, then he is not healthy.*	Man, 65 years
	As head of the household and the main provider, I am responsible for the health and well being of every single member in my house. If I fail to provide food and a good place for us to lay our heads, then I am subjecting everyone to ill health. It is my responsibility to keep the whole family healthy.	Man, 46 years
	Health for me is the ability to take good care of my children and be able to meet their day-to day needs. When this is done, then I worry less, I am able to sleep and I am also healthy.	Woman, 26 years
Linked to farm and good agricultural practices	*Fertile soil means more food and ultimately good health. Our health depends on that of our farms. Good health does not mean taking care of only our bodies, it also means taking good care of our farms, that is using good farming practices, replacing the trees we cut, and preserving the fertility of our soils.*	Man, 32 years
Emotional responsibilities	*As a woman, health is not only the absence of disease. I may look physically fit, but the fact that I am worried continuously, about how to feed my children, or send them to school makes me sick. So, although I may look physically healthy, I may in fact be dying.*	Woman, 32 years
Societal expectations	*Health! [laughing] Why are we even talking about health when we women can never say "I am not feeling well". Technically speaking, I have never experienced ill health and so cannot even describe what it feels like. I continue with my daily chores even when I don't feel well. If I allow the sickness to put me down, who will take care of my family? As a woman, I have no right to become sick.*	Woman, 42 years

definition of health, they are not translated into practice: the biomedical model still prevails. In addition, it is important to give prime consideration to local people's subjective views, knowledge and belief systems as valid ways of knowing, and to encourage health promotion strategies to embrace holistic, gender and community-relevant perspectives of health, and not just focus on treating specific diseases.

---

[4]Previously published in Dakubo, C·(2004: 54).

6.5.2 Indicators of Good Health and Poor Health

Having understood how community members conceived health and poor health, participants were asked to describe what they considered to be indicators of good health and poor health. Just like their conceptions of health, health indicators were never expressed in biomedical terms of specific diseases and morbidity rates. Instead participants expressed indicators of good health as having sufficient food to feed the family, peace in the family and community at large, high literacy levels, and unity and love among family members. Similarly, participants identified indicators of poor health to include social dysfunction, food insecurity, and lack of access to health-enhancing resources. There seemed to be similarities between men and women's views of indicators of good and poor health. Some of these views are presented in Table 6.2.

Table 6.2 Indicators of good health and poor health[5]

Indicators of good health	Indicators of poor health
Love, unity and peace among family members	Lack of unity, love, and peace among family members
Healthy-looking (plump) family members	Household food insecurity
Participation in community activities	Pride and social isolation
Low child morbidity and mortality rates	Frequent visits to the clinic
Access to education	Ignorance
Well-educated children	Childlessness, insecure old age
Access to, and control of land resources	Lack of access to farmlands
Access to, and control of proceeds of labour Financial self-sufficiency	Lack of access to income opportunities
Happy, cheerful and never worried about food, money, or shelter	Continuously worried about how to meet life's basic needs

The indicators of health identified by participants raised questions about how meaningful medical indicators, especially quantitative indicators are for assessing rural community health. Participants' indicators of health focused on issues of community cohesion, peace and unity among family members, and access to resources. Such qualitative indicators are meaningful to community members and appropriately capture their views of health and ill health, as opposed to the use of only quantitative indicators such as infant mortality rates, frequency of disease incidence, average life expectancy, among others. This is not to discount the usefulness of these indicators for health sector planning, although it is probably helpful to incorporate health indicators that are meaningful to community members for effective health promotion.

---

[5]Previously published in Dakubo, C (2004: 55).

6.5.3 Identifying Major Community Health Problems

The major health problems facing the study community were identified from two sources: one from the community clinic and the other from participants. Data on the top ten health problems for a 6-year period, 1995–2000, was gathered from the local community clinic. These diseases included: *malaria, diarrhoea and gastro-intestinal tract infections, upper respiratory tract infection (URTI) and pneumonia, skin diseases, pregnancy related complications, accidents, acute eye infections, and other gynaecological conditions*. Among the top ten diseases, five diseases: *malaria, diarrhoeal diseases, URTI, skin diseases, and acute eye infection* have consistently been ranked among the top five diseases in the community, and all seem to be influenced by environmental conditions or factors. For example, the incidence of malaria is closely associated with various land use practices and standing water bodies around the household; diarrhoeal diseases are linked to poor water quality and inadequate sanitary conditions; upper respiratory tract infection is associated with indoor air quality; while skin and acute eye infections are possibly related to contact with contaminated water sources.

To complement data from the community clinic, participants were asked to identify and rank what they perceived to be the major health problems in their community. Their responses are reported in Table 6.3.

Table 6.3 Men and Women's perceptions of major community health problems

Ranking	Men	Women
1	Malaria	Malaria
2	Diarrhoeal diseases	Diarrhoeal diseases
3	Measles	Unknown illnesses
4	Cerebrospinal Meningitis	Hernia
5	Hernia	Vision problems
6	Eye infections	Epilepsy
7	Farm-related accidents	Cerebrospinal Meningitis
8	Tuberculosis	Respiratory diseases
9	Kwashiorkor	Work-related injuries
10	Guinea worms	Rheumatism

Consistent with the top ten diseases reported by the ministry of health, both men and women's groups ranked malaria and diarrhoeal diseases as the top two health problems in the community. In the words of one man,

> malaria has become part of the community, and we have welcomed it whole-heartedly. All we need to do is learn to live with it

According to data from the clinic malaria is responsible for over half of the community's outpatient attendance, and community members have a good understanding of the disease, and its signs and symptoms. What seems to be lacking is how to effectively prevent its onset, especially in the rainy season. This is the frustration that was alluded to by the earlier quote of accommodating the disease and "learning to live with it." Diarrhoeal diseases are also widespread and affect mainly children. From a country perspective, diarrhoeal diseases are the third most common cause of out-patient attendance in most health institutions, occurring at an annual rate in children under five of 4.5 episodes per child per year, and totaling up to 10 million episodes per year in this age group (Ghana Health Survey 2003). Participants also identified measles, kwashiorkor (nutrient deficiency related), eye infections and respiratory diseases as being among the top ten diseases, and mainly affect children as well.

Men identified their most common health problems to include hernia, work-related accidents, and tuberculosis. Tuberculosis became widespread when men from the community migrated to Southern parts of the country to work as miners, especial in the gold fields. Due to poor living conditions many returned home infected with tuberculosis. Women participants identified their main health problems to include vision problems, which is occupationally related to exposure to intense fire during pot-making and charcoal burning. Other identified health problems included hernia, work-related injuries, and rheumatism, often described as joint and bone pains. Most women discussed experiencing a combination of symptoms and feelings of ill health which they were unable to categorize (listed as "unknown illness").

Overall participants seemed to have a good understanding of the major health problems in their community. However, except for malaria and diarrhoeal diseases, many other health problems identified by community members as major health problems failed to make it into the list from the community clinic, partly because they were psychosocial or symptomatic in nature. As illustrated through the discussions, these health problems are often considered illegitimate, minor and unimportant, and so have no room in the conventional health system. For example, the women's group identified "unknown" illnesses as a third major health concern with symptoms including "the inability to sleep", "worrying too much", and "thinking about our children". Such health concerns do not manifest themselves physically, they are often considered less significant both by most women experiencing them and by most health care professionals. They are often seen as complaints of a "lazy woman", and so women are either forced to conceal or present them as symptoms of legitimate health problems such as malaria, so as to receive some medical attention. While these "unknown" health problems may not be taken seriously, they certainly constitute a major psychosocial health problem, preventing women from undertaking their normal duties in an effective way. Participants' responses about their health problems reveal the need to broaden the frame of analysis of health problems, especially in developing country contexts, and in small, rural and remote communities, to include psychosocial and cultural dimensions of health. This broad view will facilitate health to be improved from a holistic perspective and not just the treatment of diseases and symptoms.

The next set of focus group discussions was designed to determine what community members thought were the driving factors behind the health problems in the community, as well as some explanations for the persistence of poor health status in the community.

6.5.4 Examining the Underlying Causes of Poor Health in the Community

A series of focus group discussions were held to explore participants' perceptions of the underlying causes of health problems facing the community. The discussions were broad with participants examining how social, political, and economic factors influenced their interaction with their surrounding environments and how such interactions impacted their health. The discussions explored issues related to rural marginalization, the impact of climate change or variability on agricultural production and food security, poverty, poor land use practices, limited access to health care services, impact of globalization and structural adjustment policies on agricultural production and health care, and land use conflicts in the community.

For example, during group discussions, participants observed that gradual climatic and ecological changes over the years had adversely impacted agricultural productivity in the community. Using timelines and important events, participants noted that a number of changes relating to weather patterns, soil fertility, vegetation cover, wildlife, and biodiversity, had taken place over the past few decades, and this had substantially impacted their farming practices and other ecological activities. For example, participants observed that rainfall patterns had become erratic and unreliable, and drought had become prevalent, affecting cropping patterns and food productivity. Participants recalled a drought in 1983, which they referred to as one of the worst droughts in the country's history. Coincidentally, this drought occurred in the early years of the implementation of the World Bank and International Monetary Fund's (IMF) structural adjustment programs, when the adverse impacts of structural adjustment policies were just beginning to unfold (Dei 1993). The removal of food subsidies, compounded by the drought rendered many households food insecure. Food and grain stamps were instituted to be issued to communities to access food, but many of these stamps failed to make it to most rural communities. As a result, many northern households replaced nutrient-rich diets, such as beans and millet with nutrient-poor, starch-based meals such as *gaari* and *konkonte*.

At the time of the study, participants discussed an internal land use conflict that had emerged between a road-building contractor and members of the community, that had both environmental and health implications. According to participants, the building contractor negotiated for community land from a few community elites, who then pressured landlords to give up portions of their farmlands for the extraction of gravel to build an airstrip in the nearby regional capital town. In exchange, landlords who complied would receive building materials, such as roofing sheets, cement and building boards. A few landlords complied and large tracts of farmlands were excavated and rendered uncultivable. Fertile farmlands were replaced with pits, trenches, and dugouts that collected water, and became fertile grounds for mosquitoes to breed and other diseases vectors to proliferate. Community members also used the standing water for bathing and for washing clothes. Participants expressed grave concern about the health implications of the situation. Staff from the community clinic observed an increase in the incidence of malaria and acute eye infections, although they could not confirm whether or not the increased incidence of the two health conditions was linked to the use of water collected in the dugouts.

This discussion led participants to further explore the various ways in which water sources become contaminated and how this adversely impacts their health. Many participants were able to make associations between the use and consumption of contaminated water and the occurrence of diseases such as diarrhoea, bilharzia, guinea worm, and skin and eye infections. Participants agreed that the lack of waste management facilities and limited sanitary facilities resulted in poor waste management and disposal practices in the community. These unsanitary practices eventually pollute fresh water bodies and water in the community dam, which is the main source of water in the community.

In addition, participants observed that with growing urbanization of the neighbouring regional capital town, community suburbs were increasingly being used as garbage disposal grounds and expressed concerned about the potential health risks of such practices. While relatively few scientific studies have been conducted on the adverse health effects of waste dumps and landfill sites, a study in five European countries determined that living near a landfill can raise the risk of having a child with birth defects, such as Downs Syndrome, by as much as 40% (Vrijheid 2000).

Participants in the various focus groups also explored how their daily interaction with the biophysical environment could adversely impact the environment and also pose threats to their health and well-being. One major activity identified by participants was the conversion of virgin or long fallow lands to farmlands or for settlement purposes:

"As a young man of age and recently married, I need to step aside and build my own little place for my family."

Such extensive land clearing often involves clear cutting and bush burning, while leaving behind a few economic and fruit trees. The conversion of vast tracts of virgin land for agriculture or settlement purposes has been associated with the emergence of new diseases as was seen in the case of Lyme disease in Connecticut (Levins et al. 1994; Schrag and Wiener 1995). While participants acknowledged that such land use practices were ecologically destructive and might adversely affect health through the loss of medicinal plants, game and other wild food sources, participants found it difficult to make any direct association between the conversion of virgin land and the proliferation of disease vectors. Drawing on prior knowledge and from literature, health and agricultural experts in the research team explained how extensive virgin land conversions sometimes eliminated disease vector predators and competitors, creating opportunities for new species to colonize the area. When people come into contact with these environments, they stood the chance of contracting both old and new diseases (Levins et al. 1994). Despite these shortcomings, many participants were able to make linkages between how various land use practices impacted the environment and their health.

Group discussions revealed that lack of access to quality health care services was a major underlying cause of poor health in the community. According to participants, factors such as long commuting distance to the nearest hospital, limited transportation and expensive transportation costs, high health care costs and some health belief systems prevented them from accessing quality health care. In addition to these factors, women also identified lack of autonomy and decision-making powers to seek health care. There were also concerns about interrupting household organization, and limited social and family support. The high costs of health care services in district and regional hospitals prevent many rural community inhabitants from seeking health care at that level. For example, in Ghana, from the 1980s through to the 1990s, high health care costs had been blamed partially on structural adjustment programs and low government health spending. Cuts in health spending significantly affected the availability of health supplies and basic drugs. User fees were introduced for casualty and polyclinics, and other services such as, laboratory tests, drugs, medical examinations, and medical and surgical treatments (Anyinam 1989). The high costs of drugs and the implementation of user fees have persisted until today and have greatly affected the provision and utilization of health care services, especially in rural areas (Konadu-Agyemang and Takyi 2001; Nyonator and Kutzin 1999).

Finally, during group discussions, the research team observed that both men and women referred quite frequently to poverty as one of the key drivers of environmental degradation and poor health. Participants observed that they had no other source of livelihood but to "live from the land," and so while being cognizant that their activities could negatively impact the surrounding ecosystem, their options were limited. Some participants blamed the national government for neglecting rural communities and concentrating all health care resources and employment opportunities in urban and southern regions of the country. Some women expressed interest in small-scale micro-enterprises such as food processing, but cited some major constraints to include lack of access to capital or small loans, lack of electricity for storage facilities and equipment, and limited business and marketing skills. Overall, the group discussions were very informative and generated thoughtful insights and ideas that would not have emerged through the use of other methods such as one-on-one interviews.

6.6   Analysing Participants Responses from a Political Ecology of Health Perspective

In Ghana and other parts of the world, many explanations of ill health still focus on blaming the victim for engaging in poor and inappropriate hygienic health practices, behaviours, and lifestyle. They also blame them for holding onto health-deteriorating cultural belief systems, practices, and traditions that adversely impact their health, as well as failing to take the necessary precautions to stay healthy. As discussed in earlier chapters, the focus on individual-level factors fails to take into account the multiple and constraining factors that influence individual choices and rural health outcomes. Similarly, explanations for the predominance of environmental-related health problems in many rural communities focus on the irrational land use practices of locals, without situating such practices within the context of uneven power relations surrounding the use of and access to environmental resources, and how this engenders poverty and minimizes livelihood options.

A political ecology of health analytical framework challenges such atheoretical interpretations and explanations of community health problems, and instead seeks to examine them in light of broader historical and socio-political contexts. For example, during group discussions, participants made reference to poverty, rural marginalization, and unaffordable health services as some of the main factors preventing the realization of good health in the community. A thorough examination and response to these concerns require that we ask questions such as: how is it that poverty, rural marginalization and all the inequities seem to prevail and persist in this community? What role did colonial health, agricultural and other policies play in perpetuating this trend of spatial and socioeconomic inequalities? What role did past economic reforms by the World Bank and the International Monetary Fund play in shaping rural health care? How do externally-determined priorities by the World Health Organization and other international development agencies impact healthcare delivery and practices in Ghana? What role does globalization and unfair trade agreements play in influencing the rural economy?

As mentioned above, the colonial legacy provides a good starting point to begin to understand the basis for the social disparities. For example, Ghana's health system was designed to serve colonial masters who were based in coastal urban towns and ports, and so the location for many health facilities was in southern coastal towns to the neglect of northern and rural areas (Twumasi 1981; Kunfah 1996). Colonial health services focused on the elimination of tropical diseases to allow for colonial capitalist expansion. Technocentric remedies such as spraying mosquitoes with DDT, and the development of vaccines to counter diseases took precedence over preventive measures such as the provision of clean water, sanitation, and good housing. Clean water, sanitation, and good housing were provided for only colonial masters, and later the labour force, as it was deemed important to keep the labour force healthy (Aidoo 1982; Twumasi 1981). The selective location of health facilities, and provision of health services, to some extent, shaped the spatial inequities and health disparities that characterize the North and South, and between rural and urban areas of the country, as depicted in the World Bank study discussed above.

Following independence from the colonial masters, it was anticipated that health care delivery and services would be re-oriented to suit the health needs of the Ghanaian populace, and pay particular attention to preventive measures. However, the persistence of colonial class-based structures still favoured elites and so continued to organize health services to suit them rather than the ordinary citizen living in peripheral regions. Besides, it was easier to continue to maintain existing health facilities than to build new ones in other locations. In fact, the few health facilities that were built in rural areas were built by missionaries who were eager for more converts and saw health and education as two good avenues to recruiting converts.

Also, Ghana's continuous dependence on foreign capital and external expertise following independence, meant that economic development and health decisions and policies were still determined externally. With respect to health, international health organizations such as the World Health Organization (WHO) became the new centres of expertise, and their policies and programs were in tune with the colonial era (Randall 1998). For example, the WHO favoured large-scale disease eradication and immunization campaigns as these demonstrated concrete evidence in terms of numbers of people immunized, as well as justified the judicious use of donors' money. Because of the need to demonstrate results, diseases that could easily be addressed through technological solutions were considered high priority, while those that required education and the establishment of primary care facilities were low priority (ibid).

The introduction of the Alma Ata declaration in 1978 was meant to re-orient health care from a medical focus to a community-based, preventive system of care. However, Ghana and other African countries could not implement these measures, partly because, the Alma Ata Declaration was released at the time when many African countries were going through tough economic crisis. Albeit attempts to adopt and implement broad-based community health models, the curative, oriented and disease-focused biomedical model continues to dominate most public health practices in Ghana. The biomedical health model considers individuals as responsible for their health outcomes, and so the individual becomes the appropriate target for intervention (Pierce 1996; Susser and Susser 1996). Critics have pointed out that, excessive emphasis on the individual obscures the role of structural inequalities and power imbalances in perpetuating uneven patterns of health. It ignores the sociopolitical contexts in which individual decision-making about health choices and action takes place (Lee and Garvin 2003; Minkler 1994; Neubauer and Pratt 1981).

Similarly, explanations that blame peoples culture and beliefs systems for poor health outcomes fail to reckon the discursive means through which local peoples' identities and health practices are constructed to suit the ideological, political and material interests of dominant discourses. Modern medicine sees traditional values and belief systems as obstacles to rural health development. These values are perceived as irrational and self-destructive and stand in the way of modern medical practice (Farmer 2001; Harper 2004; Twumasi 1981). For example, it is difficult for Indigenous views, perceptions and practices to be seen as valid ways of knowing and be fully embraced by dominant forms of practice, although this is beginning to change with integrated healing practices.

One other constraint in how public health is practiced in rural African communities, is the tendency to perceive rural people as a homogenous group with similar health experiences. It is based on such homogenous constructions of health problems that universal health promotion strategies are developed for "rural people" or "the developing world". Some universal packages, such as maternal and child health clinics, fail to take into account the unique and varied needs and experiences of the women and their infants, and of rural people in general. During group discussions, it was apparent that how people conceptualized health, what they perceived to be indicators of good and poor health, as well as their health problems and needs all varied across gender, age, socioeconomic status and access to social support networks. For example, work-related and psychosocial health problems were widely experienced by women compared to men. And among women, younger women were more concerned about work-related health concerns, while elderly women were most concerned about social networks. Also, participants with some formal education tended to conceive of health from western perspectives, compared to others with no formal education. Thus, it is important to pay particular attention to the multiple and changing identities that individuals assume, and tailor health interventions to reflect and incorporate these multiple identities.

In addition, explanations for environmental health problems that blame people for irrational use of environmental resources or blame poor land use practices for ecosystem degradation adopt a uni-dimensional and limited view to how such activities emerge. For example, the story of the road-building contractor and his activities are exemplary in this study. The building contractor acted based on his privileged position of being in a higher socio-economic bracket than many community members and took advantage of the poverty-stricken nature of community members to offer building materials in exchange for farmlands. He also took advantage of the lax or absent environmental regulations in the community, and was fully aware he did not live in the community and so will not be affected by the resulting environment and health impacts of his activities. These issues, although at a micro-level, typifies what goes on in the broader, global scale where other prosperous regions are able to recklessly exploit the natural resource base of poorer regions, and in the process displace the environment and health costs. It is on this basis that the issue of scale is very important in analyzing health determinants from an ecosystem perspective. Also, poor land use practices need to be examined from a broader context. For example in this community land use practices can be examined from the lens of the IMF/World Bank-led structural adjustment policies (SAP), which caused the removal of subsidies on basic commodities such as food, agricultural products and fuel, and as a result increased rural poverty, household food insecurity and increased dependence on natural resources. This policy move forced the rural poor to encroach on marginal lands and fragile ecosystems, while making use of intensive and unsustainable farming practices (Alubo 1990; Kanji et al. 1991; Kessler and Van Dorp 1998). The structural adjustment measures were also implemented to increase tradables and so the cash-crop sector flourished at the expense of rural subsistence (Cornia et al. 1987; Kanji et al. 1991; Woodward 1992). The SAP favoured large-scale commercial farmers and landowners compared to small-scale subsistence farmers. For example in the Upper West Region of Ghana, large-scale cotton production was encouraged at the expense of food crop production. The cultivation of cotton made use of intensive farming practices, replacing the traditional hoe farming with mechanized farming and agrochemicals. Farmlands that were used for cotton production eventually became unsuitable for food crop production. SAPs also placed an extra burden of work on women in smallholder households through pressure to produce more crops for sale. Increased workloads combined with the inability to meet the family's consumption needs undermined the nutritional status of both women and children, and increased their susceptibility to infectious disease and other maternal and child related morbidities and mortalities (Kanji et al. 1991). As Bradley (1993) explains, nutrition plays an important role in understanding the high incidence of infectious diseases in developing countries. Low nutritional status predisposes an individual to infection, and because the immunological status of a malnourished individual is already diminished, the course of infection is more severe compared to a well-nourished person. In addition, not only are malnourished people more susceptible infectious diseases, but the infection itself further augments malnutrition, resulting in a synergistic relationship. Political ecology therefore allows for the understanding that problems of land degradation, food insecurity and malnutrition in Northern Ghana are simultaneously political-economic problems and cannot be examined outside these contexts.

6.7 Conclusion

Health problems facing rural communities around the world cannot be fully understood and improved without situating them in broader historical, political and socioeconomic contexts. The factors causing ill health are often interpreted superficially, but a good understanding of the underlying causes of community health problems requires the use of rigorous analytical frameworks. In this study, I combined the ecosystem approach to human health, with a political ecological framework, and participatory action research, infused with poststructuralist perspectives, to analyse the multiplicity of factors affecting human health concerns in a rural community in Northern Ghana. Such an integrated framework allows for a nuanced and context-based understanding of peoples health problems and their underlying causes. The framework allows the research to counter dominant, yet simplistic, explanations about the persistence of poor health in rural communities and to work with community members in finding appropriate solutions. The study also provided the opportunity for participants to articulate their own views of health, what they perceived to be the major factors causing poor health in the community, and evaluated their role in responding to these factors. In contrast to conventional health research approaches, this study allowed participants to articulate their health concerns and examine them in ways that were meaningful to them, both culturally and socially, and not externally imposed.

In the next chapter, we will take this study further, and examine how this first phase of the project laid the foundation for participants to engage in a strategic planning exercise to develop a plan for building a healthy community.

References

Aidoo TA (1982) Rural health under colonialism and neocolonialism: a survey of the Ghanaian experience. Int J Health Serv 12:637–657

Alubo SO (1990) Debt crisis, health and health services in Africa. Soc Sci Med 31:639–648

Anyinam CA (1989) The social costs of the international monetary fund's adjustment programs for poverty: the case of health care development in Ghana. Int J Health Serv 19:531–547

Bradley DJ (1993). Environmental and health problems of developing countries. In: Environmental and human health. Wiley, Chichester, pp 234–246. Ciba Foundation Symposium 175

Cornia G, Jolly R, Stewart F (1987) Adjustment with a human face. Clarendon, Oxford, UK

Dakubo C (2004) Ecosystem approach to community health planning in Ghana. EcoHealth1: 50–59

Dei GJS (1993) Learning in the time of structural adjustment: the Ghanaian experience. Can Int Educ 22(1):43–65

Farmer P (2001) Infections and inequalities: the modern plagues. Updated Edition. University of California Press, Berkeley, CA

Ghana Living Standards Survey (2000) Ghana statistical service. Ghana Accra, Government of Ghana

Ghana Statistical Service (GSS) and Macro International Inc (1999) Ghana Demographic and health survey. GSS and MI, Calverton, MD

Ghana Statistical Service (GSS) and Macro International Inc (2003) Ghana demographic and health survey. GSS and MI, Calverton, MD

Ghana Statistical Service (GSS) and Macro International Inc (2008) Ghana demographic and health survey. GSS and MI, Calverton, MD

Government of Ghana (2008) Budget and Economic Policy Statement of Ghana, Ministry of Finance and Economic Planning, Accra, Ghana

Harper J (2004) Breathless in Houston: a political ecology of health approach to understanding environmental health concerns. Med Anthropol 23(4):295–326

Kanji N, Kanji N, Manji F (1991) From development to sustained crisis: structural adjustment, equity and health. Soc Sci Med 33:985–993

Kessler JJ, Van Dorp M (1998) Structural adjustment and the environment: the need for an analytical methodology. Ecol Econ 27:267–281

Konadu-Agyemang K, Takyi BK (2001) Structural Adjustment Programs and the political economy of development and underdevelopment in Ghana. In: Konadu-Agyemang K (ed) IMF and World Bank sponsored structural adjustment programs in Africa: Ghana's experience. 1983–1999 Ashgate Publishing Company, Burlington, VT, pp 17–40

Kunfaa EY (1996). Sustainable rural health services through community-based organisations. Spring Research Series No.16, University of Dortmund. Dortmund: SRING Centre, 1996; p 25, 54

Lee RG, Garvin T (2003) Moving from information transfer to information exchange in health and health care. Soc Sci Med 56:449–464

Levins R et al (1994) The emergence of new diseases. Am Sci 82:52–60

Maguire P (1996) Considering more feminist participatory research: What's congruency got to do with it?. Qual Inq 2:106–108

Minkler M (1994) Challenges for health promotion in the 1990s: social inequities, empowerment, negative consequences, and the common good. Am J Health Promot. 8(6):403–413

Neubauer D, Pratt R (1981) The second public health revolution: a critical appraisal. J Health Polit Pol Law 6(2):205–228

Nyonator F, Kutzin J (1999) Health for some? The effects of user fees in the Volta Region of Ghana. Health Policy Plan 14:329–341

Pierce N (1996) Traditional epidemiology, modern epidemiology and public health. Am J Public Health 86(5):678–683

Randall P (1998). Health care systems in Africa: Patterns and Prospects. Report from the workshop, health systems and health care: patterns and perspectives. 27–29 April 1998. The North-South Co-ordination Group. University of Copenhagan and The ENRCA Health Network

Schrag SJ, Wiener P (1995) Emerging infectious diseases: what are the relative roles of ecology and evolution?. Trends Ecol Evol 10:319–324

Songsore J (1983) Intraregional and interregional labour migrations in historical perspective: the case of north-western Ghana. University of Port Harcourt, Faculty of Social Sciences, Port Harcourt, Nigeria

Susser M, Susser E (1996) Choosing a future for epidemiology: eras and paradigms. Am J Public Health 86:668–673

Twumasi P (1981) Colonialism and international health: a study in social change in Ghana. Soc Sci Med [Med Anthropol] 15B(2):147–151

Vrijheid M (2000) Health effects of residence near hazardous waste landfill sites: a review of epidemiologic literature. Environ Health Perspect 108(Suppl 1)

Wallerstein N (1999) Power between evaluator and community: research relationships within New Mexico's healthier communities. Social Sci Med 49(1):39–53

World Bank (2003) Ghana-Second Health Program Support Project. Project Appraisal Document. Project # 24842-GH, Washington D.C.

Woodward D (1992) Debt, adjustment and poverty in developing countries. Frances Pinter in association with Save the Children, London

Chapter 7
Planning for a Healthy Community: A Case Study – Phase II

Contents

7.1 Introduction

This chapter builds on the first phase of the ecohealth research project that was described in the previous chapter. This chapter presents the findings of a strategic planning process (SPP) that was organized with both men's and women's groups to map out their vision of a healthy community. The strategic planning process took community members perceptions of health and their articulation of the factors driving poor health in the community as the basis from which to plan for a healthy community. The process proceeded through a series of community workshops in which the research team used focus group discussions to guide participants through the various phases of the strategic planning process.

As discussed in the methods chapter, Chapter 5, the strategic planning process is an approach that allows participants to look beyond their immediate problems and to recast their abilities and efforts in identifying and implementing strategies to

C.Y. Dakubo, *Ecosystems and Human Health*, DOI 10.1007/978-1-4419-0206-1_7,
© Springer Science+Business Media, LLC 2011

achieve a desirable future. The process leads discussions through a series of logical steps that allow participants to first articulate a vision of what their perceptions of a healthy community looks like. They then proceed to identify the obstacles preventing them from achieving that vision. Next, they articulate what it would take (resources, abilities) to achieve their vision, and finally, they devise a plan of action to implement the necessary actions. This process builds on the principles of participatory action research and helps communities develop planning capacities to meet their goals among diverse stakeholders.

To set the stage for the process, the research team took participants through a guided imagery of what a healthy community could look like 5 years into the future; a future that is rid of all major health problems and their associated environmental and social causes identified in earlier focus group discussions. To do this, participants were asked to respond to a focus question identifying the attributes, features and conditions they would like to experience in this new healthy community. The research team developed similar focus questions for the other steps of the strategic planning process (Fig. 7.1). Participants seemed enthusiastic about the process and were optimistic about the prospects of building a healthy and sustainable community for themselves and for their children.

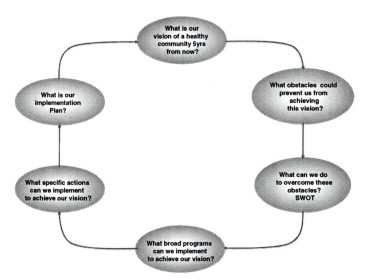

Fig. 7.1 Planning for a healthy community

With this enthusiasm also came a number of challenges. One was achieving consensus among a diverse group of people with different and conflicting needs. In group processes, there is the tendency for the louder voices to be heard, and those that need to be heard silenced. Hence, to what extent will the vision and plan accommodate the needs of both men and women, children and adults, young and old, employed and unemployed, rich and poor? To what extent can these diverse groups work together to make the plan a healthy one? These are questions the research team grappled with and tried to use the strategic planning process as a tool to encourage these diverse views to emerge and be thoroughly deliberated upon, before participants made a final resolve on how they wanted to proceed.

Another challenge related to re-orienting community members mindset from blaming the government to embracing a spirit of self-help, and focusing primarily on their strengths and resources; asking questions such as what can we do for ourselves versus what can they do for us? Such a perspective allows community members to plan and act based on immediacy and capability, while organizing to lobby the government to support their plans. It also allows them to gain a good understanding of their problems and become better positioned to confront issues of political marginalization and rural neglect. In general, the planning process fosters critical self-examination, community competence, and self-reliance.

Finally, effectively facilitating focus group discussions and connecting information between the various phases of the strategic planning process requires good facilitation skills. Despite the initial training of members of the research team on some group facilitation techniques, it was still challenging for the team to accommodate varied opinions in an effective and organized way. It was also very difficult for members of the research team to hold back their own biases and suggestions, while encouraging those of participants to emerge in a non-threatening atmosphere.

In presenting the discussions of these focus group discussions, I will make use of direct quotations where necessary. In certain instances, however, participants' responses are summarized into appropriate themes and presented in tabular forms.

7.2 Mapping Out A Vision for a Healthy Community

The vision workshop is a process of mapping out a shared picture of that ideal place that participants would want to live in, work and play. It requires participants to expand their thinking beyond current obstacles and to identify inspirational and far-reaching, yet practical ideas for the future. The attributes for this ideal community represent community members' values, priorities, and desired needs, and how they hope to function as a community in the future. The visioning exercise, if carried out properly, can provide participants with a common purpose and shared commitment to work towards achieving their desired goal. People will only own the outcome of a vision if it is a vision that they participated in creating and not one that is externally imposed.

The group began the vision workshop by recapping the major themes that emerged in earlier focus group discussions (discussed in previous chapter), including their perceptions of health and what they perceived to be indicators of good health and poor health; and their perceptions of the major health problems in the community and their associated causes. Participants were encouraged to envision a healthy and sustainable community from a holistic perspective, as they had done before, taking into consideration aspects of the natural environment, the built environment, social, cultural and political systems, and aspects of the rural economy.

In the men's group, their vision elements emphasized issues related to the vitality of the biophysical environment. They identified attributes such as fertile soil, good grazing pastures, and woodlands as essential features of a healthy community. They were also concerned about the equitable distribution of resources, both internally and at the national level. They also identified the need for clean water supply, roads, electricity, waste management services, and a better working relationship with local and regional government departments.

The desired elements of women, among other things, focused on peace and unity among family members, sense of community and belonging, availability of social support networks, and access to better education for girls.

The vision elements identified by both men and women were categorized into seven major vision themes, including:

- A viable natural environment – healthy ecosystems
- Improved health status and good quality of life for community members
- An equitable and supportive social environment
- A vital and sustainable rural economy
- Increased access to basic services and infrastructure
- Increased public participation in community activities
- Development of human capital

A detailed list of the vision elements identified by men and women's groups are presented in Table 7.1.

Both men and women made strong linkages among ecosystem health, human health and sustainable community development. According to most participants, a community with a sustainable natural resource base, productive farmlands, sustainable agriculture, grazing lands and clean fresh water bodies, is more likely to support a healthy population than a community with depleted natural resources and a degraded environment. One middle-aged man put it this way:

> From folklore and stories from our grandparents, we know that in the olden days the land was more productive than it is today. Our ancestors did not struggle to find food or meat. The land was productive and everything was plentiful. I understand people rarely became sick in those days, and even when they did they had all the medicine at their backyard. These days our lands are bare and non-productive, food has become so scarce and we are faced with very strange health problems. The forestry and agricultural officers blame us for the current state of our land, but it is not our fault. We have no other alternative but to continue to live off the land. So for me, I think if we can find ways to take good care of our land, then we may be able to live as healthily as our ancestors.

Some women expressed a desire for a community with a strong sense of unity and interconnectedness; a community that embraces communal spirit, provides social support and is able to cater to the varying needs of its members. In one young woman's words,

> ...unity drives every successful event. If there is no unity among us, both at home and in the community, then all our efforts will be in vain.

Table 7.1 Vision elements of a healthy community

Vision elements	Men (M)/ Women (W)
Clean and healthy surrounding environments	M/W
Improved health status for all community members	M/W
Productive and sustainable agriculture	M/W
Peace and unity among family members	W
High levels of school enrolment and education in the community	W
Equitable and accessible distribution of health services	M/W
Equitable resource distribution within the community and among community members	M/W
Ability to meet basic needs (food, water, sanitation, proper nutrition, shelter, income)	M/W
Access to basic infrastructure (roads, communication services, electricity, waste management)	M
Access to markets, a growing community economy, and financial self-sufficiency	M/W
Active community involvement in all phases of community development programs	M/W
Healthy and sustainable ecosystems (farmlands, water bodies, rangelands, woodlands)	M/W
Greater awareness of health promotion strategies and behaviours	M/W
A sense of community and belonging, and available social support networks	W
Egalitarian relationship with local government departments	M
Equitable distribution of government resources	M

There were concerns among participants on how to reach beyond their differences to achieve communal goals. Most of these differences were enhanced by the unequal allocation of resources flowing from "development" projects in the community. Participants expressed disappointment at the "greed" and "selfish" attitudes of some community members, as they manipulated international development projects such as woodlots, tree nurseries and plantations to be established in the sections where they lived, to the detriment of other sections in the community. Participants also referred to some individuals as "saboteurs" who deliberately set fire to community woodlots. The men's group interpreted such acts of arson as acts of resistance or protest against the concentration of international development projects in one section of the community. With respect to community functioning, both groups envisioned a community in which they would participate actively in the planning and decision-making of community development projects. According to participants, for so many decades they have played passive roles in activities influencing their lives, partly because they have always lacked confidence in their ability to be influential, and most often than not, denied the opportunity to be active players in their own development efforts. They see a new community in which strong and collaborative partnerships would be established with local government, the private sector and non-governmental organizations.

Participants envisioned a productive rural economy and improved socioeconomic conditions for all members of the community. With agriculture being the mainstay of the community, many men envisioned increased access to agricultural loans to enable them buy the necessary equipment to help bolster agricultural productivity. They indicated that access to value-added processing equipment, and storage facilities, as well as increased technical assistance from the agricultural department would help enhance food security in the community and increase the start-up of micro-enterprises. Other income-generating activities, such as pot making by women and carving by men, could generate a considerable amount of income for community members if these industries were promoted and supported nationally and internationally. Improved socioeconomic conditions would then allow people to access basic services such as food, water, good housing and education.

During the visioning exercise, the research team was mindful of raising undue expectations among participants. Participants were often reminded to identify activities that were practical and could easily be implemented with minimal external help. Activities that required external help were not discouraged but only placed at a lower level of priority. Members of the research team, especially those from government departments, were particularly mindful of unnecessary promises, as these, if not fulfilled, could strain relations with the community after the research was over.

Despite a few disagreements, both men and women articulated a vision that depicted a holistic and ecological perspective of health, giving equal importance to environmental, social and economic issues. The vision mapped out by participants incorporates some of the principles of a healthy community outlined by the WHO Healthy Cities/Communities Initiative. It reflects some of the attributes identified by Hancock's (1993) model of a community ecosystem, discussed in earlier chapters. This model identified six attributes of a healthy and sustainable community, including *environmental viability, ecological sustainability*, *livability*, *community conviviality*, *social equity* and *economic prosperity*. Hancock observed that these attributes of a healthy community cannot be translated into action without two key drivers of the change process, education and governance, which comprise of knowledge development, awareness creation, empowerment, participation, and performance by government.

While participants were delighted and inspired by the vision they had outlined for themselves, they were also concerned about the roadblocks, barriers and obstacles that stood in their way to achieving this vision of a healthy community.

7.3 Identifying Obstacles to Achieving a Healthy Community

Participants were asked to identify what they perceived to be the major obstacles or barriers that could prevent them from realizing the vision they had outlined in the previous session. The barriers identified by participants were categorized into four broad themes: geographic, institutional, socioeconomic, and behavioural barriers (Table 7.2).

Table 7.2   Barriers to realizing the vision of a healthy community

Theme	Examples of barriers
Geographic/Climatic	Changing climatic conditions and impact on ecosystems
	Loss of restorative ability of natural environments
	Rural isolation, neglect and marginalization
Institutional	Lack of fit between community needs and environmental health programs
	Poor intersectoral coordination
	Poor health and environmental communication strategies
	Lack of strict environmental regulations
Socioeconomic	Poverty
	Poor social support networks
	Low status of women
	Lack of community participation in health planning
	Lack of access to health care planning
Behavioral	Cultural, religious beliefs and practices
	Lack of participation in health promotion programs
	Lack of personal initiative for preventive health care
	Non-utilization of community health services
	Non-compliance to adopt sustainable environmental practices
	Ignorance of some health risks and causal pathways

The geographic and climatic obstacles identified by participants were similar to those discussed in the previous chapter. These included the naturally harsh ecological conditions, low soil fertility, climatic variability, and gradual environmental degradation. Participants acknowledged that while some of these ecological conditions were natural and could not be altered, environmental degradation was partly due to their poor socioeconomic status as a result of rural marginalization, and so addressing these structural obstacles was highly prioritized.

Participants also identified a number of institutional factors constraining the realization of their vision. Significant among them was the difficulty in ensuring that environment and health programming suited community needs. They also identified poor coordination among local government departments, ineffective communication strategies by field workers, and lack of effective legislation, especially in the environment sector as some key factors. As discussed previously, many rural health services are operated based on externally determined health priorities and rarely address pressing health problems in the community. Although Ghana's Ministry of Health has pledged to pay more attention and devote more resources to rural areas, such as rural areas receiving the greater part of increases in health service budgetary allocation, this has rarely been translated into action. Also, while community participation in health programs is in vogue, participants observed that they rarely had any voice in community health related planning.

Given the holistic perspective of health articulated by participants in earlier discussions, achieving a healthy and sustainable community requires intersectoral collaboration among government departments, who in turn will have to work with community members, the private sector and all stakeholders in planning a healthy community. However, participating government officials observed that such inter-departmental collaboration is fraught with difficulty because of the varying and sometimes competing departmental mandates and goals. For example, which department will be responsible for providing resources for such joint ventures? How will departments justify the allocation of resources to such projects? How will they protect their departmental interests while serving others? Some male participants observed that piece-meal and sectoral approaches to rural intervention, in general, is ineffective and time consuming as local people have to attend separate meetings organized by the various departments.

Participants also observed that creating a healthy community goes beyond inter-sectoral collaboration, but also involves other stakeholders, such as the private sector whose activities affect community health one way or another. They observed that while it may be relatively easy to bring local government departments together, it is difficult to bring to the table individuals and private businesses who perceive the community as a "resource depot", and care very little about people's health or the damage they may cause to the environment.

In the women's group, some female participants identified the erosion of social support networks as major barriers to improving community health. Many women viewed support from families, friends and communities as necessary for good health. Such social support networks could be very important in helping people solve problems and deal with adversity. They could also serve as pathways for informal learning, and help people maintain a sense of control over their life circumstances. Accordingly, these support networks in the community have begun to deteriorate as a result of economic hardships. In recent times, traditional support systems are being eroded as people tend to rely on community organizations, such as churches, youth groups, trade unions and other political bodies, of which they are part. In the event that one does not belong to any of these groups, then community or social support is often lacking and people find themselves lonely and anxious. In one woman's view, "we face problems everyday, and when there is no one to turn to for help, it not only affects you mentally, but it affects you physically and your ability to make good decisions." Some women observed that because of time constraints, restrictions from their husbands, and religious beliefs, not every one can belong to a community organization and so formal support groups could be organized to accommodate all women and provide them with peace of mind.

In addition to the structural barriers outlined above, participants shifted the focus to themselves, examining how their attitudes, behaviours and beliefs could constrain their ability to engage in appropriate health practices. The most common barrier identified by both men and women was limited time to engage in preventive and health-enhancing practices. Most participants indicated that they were preoccupied with securing basic needs such as food, water, shelter and money that they placed little value in protecting their health, or in participating in environment and health education programs. Women, on the other hand, explained their health was secondary to their children's health and are often unable to invest time in their own health. As observed by one young lady:

Sometimes, it is not the lack of knowledge on our part to protect our health, it is rather a question of whose health is more important. The children easily fall sick and so we do all we can to protect their health.

Another barrier that was identified in the men's group was the continuous denial of men as vulnerable to ill health. Many men see illness as a sign of weakness, and so they tend to ignore any potential signs and symptoms of ill health. They find it difficult to discuss health problems with female community health nurses and are often apprehensive about how others will perceive them. According to the community health nurse, this attitude by men is reflected in the low male patronage in the village clinic.

Despite occasional disagreements, participants observed that adopting positive health behaviours, attitudes and practices would move them a step closer to achieving their vision of good health. They emphasized the need to intensify environmental health education and making use of more culturally appropriate communication strategies.

7.4 Assessing Our Strengths and Resources

Contrary to seeing communities as passive victims needing external assistance, there is now growing emphasis to look at communities as having assets, strengths, resources and capabilities that can be harnessed to improve their living conditions (Kretzmann and McKnight 1993; McKnight 1985). Communities possess a wealth of knowledge and experience, values, norms, and other locally available assets that can play a major role in efforts aimed at improving their livelihood.

During group discussions the research team asked participants to identify what they perceived as their strengths and resources that could help overcome the barriers identified and facilitate the achievement of their vision. This step allows community members to focus on their own strengths and resources, relying less on external assistance. The issues identified during this exercise are presented in Table 7.3.

Table 7.3 Perceived strengths and assets of community

Strengths
A strong sense of community coherence
Possession of valuable traditional environmental and health knowledge
Well-organized community groups (women, youth and religious organizations)
On-going international development projects
Dynamic chief and community leaders
Receptive to external development programs
Good working relationship with government extension workers
Enthusiastic and hardworking community members
Increasing number of highly educated youth
Strong ability to mobilize and work on community projects

Good community organizing emerged as their strongest asset. The village has a dynamic chief, who is backed by able elders, and has always been supportive of community development projects. Over the years, he has committed resources, such as land, to community projects and has also taken keen interest in research activities in the community. Also, participants identified strong traditional knowledge on environment and health issues as a key asset, which could be harnessed by extension and other local government workers through the development of professional/lay knowledge synergies, rather than the "expert-know-all" mentality that elides local people's knowledge and creativity.

According to participants, the community has a growing number of highly educated youth, some of who hold top national government positions. These highly educated youth have always taken keen interest in the development of their community, supporting it financially and otherwise to undertake development projects. They are also in key positions to lobby the government to allocate resources for the establishment of some basic infrastructure and services. During the research process, there were efforts to connect the community to the national electricity grid, a move seen by community members as a big step that could facilitate the establishment of small-scale enterprises and value-added agricultural processing facilities. Participants were confident they could harness these assets to help overcome some of the structural barriers, and put them on the path to achieving their vision of a healthy community.

7.5 Identifying Strategic Directions and Systematic Actions for a Healthy Community

After identifying the barriers, strengths and resources to achieving a healthy and sustainable community, participants began to explore some broad strategies, followed by systematic actions that could lead them to their desired future. This section combined two stages of the strategic planning process: the strategic directions and the systematic actions. For the strategic directions phase, participants were asked to identify possible strategies or broad proposals, that when implemented would help them achieve their vision of a healthy community. These strategies were then categorized into broader themes for which specific actions were identified. The research team urged participants to be creative and bold in identifying their strategies and actions. The team encouraged participants to identify strategies and actions that were practical and attainable, and within the means and capabilities of the community. This was done to encourage participants to develop some confidence in their own capabilities of finding solutions to community problems. It was also meant to avoid the trap that many rural communities fall into when they suggest strategies that cannot easily be achieved by the community itself, forcing them to rely on external assistance, or assistance from government and non-governmental organizations (Bergdall 1993). Contrary to the magic of external assistance, it has been observed that except for actions requiring policy backing, many actions needed to improve human health especially through primary environmental care can be achieved through local efforts (World Health Organization 1994). Hence, participants were encouraged to envision the realization of their healthy community as primarily dependent on them, their resources and strengths, and the partnerships they have formed with their allies, both within and outside the community.

At the end of the group workshops, participants identified six broad strategic directions, including:

- Managing our environment in ways to promote health and to prevent disease
- Initiating intersectoral approaches to community planning and development
- Encouraging the use of more inclusive and participatory approaches for community planning
- Strengthening environment and health education
- Building on local knowledge systems, and indigenous strengths
- Building community capital

7.5.1 Building Community Capacity to Manage Local Environments Sustainably

Participants observed that since major community health problems such as malaria, diarrhoea and acute respiratory infections are environmentally-mediated, it was important for community members to pay greater attention to managing environmental conditions in ways that will prevent the occurrence of these diseases. The home environment also needed to be managed so as to effectively respond to problems such as poor water quality, food contamination, inappropriate waste and excreta disposal, crowded and poorly ventilated living spaces, and smoky indoor air pollution so as to reduce the occurrence of associated health problems.

While participants indicated that they had very little control over climatic conditions and natural environmental phenomena, there were certain practices they could begin to undertake that would prevent further environmental degradation. Some of these practices identified by the men's group included the need to establish local legislation and community by-laws, and form enforcement committees to prevent the use of bush fires for trapping game animals during the hunting season, and to levy fines on people felling trees indiscriminately, especially on fragile landscapes. Both men and women's groups expressed interests in working with local nongovernmental organizations and extension officers to undertake activities such as tree planting, watershed and rangeland management, vegetable gardening, recycling and composting domestic wastes, and the construction of more sanitary facilities in the community. Participants thought that the community already had a head-start on such projects as a result of on-going Canadian International Development Agency activities in the community. The only problem they observed was the waning support and participation in the project, especially from some sections of the community that benefited least from initial development projects.

Table 7.4 Proposed actions to prevent the occurrence of diseases

Health problem	Actions
Malaria	Reduce/eliminate breeding sites for mosquitoes
	Use appropriate farming practices
	Proper maintenance of water supplies
	Screen doors and windows
	Self-protection from mosquitoes
Diarrhoea	Proper maintenance of water supplies
	Practice good hygiene and sanitation
	Reduce breeding sites for flies
	Protect food and drinking water from "dirt"
	Reduce waste production
	Cook meals thoroughly
ARIs	Build highly ventilated houses
	Use efficient household stoves
	Keep children away from smoky kitchens
	Cook in yards (open areas)

Participants suggested some practical actions that could be implemented at the community and household levels to prevent and/or reduce the incidence of malaria, diarrhoea and acute respiratory infections (ARIs). These actions, listed in Table 7.4 range from taking basic precautions to avoiding mosquito bites to building well-ventilated houses.

Some women cautioned that identifying these actions without making the conscious effort to implement them was meaningless. Members of the research team also pointed out that, in addition to these actions, it was important to continue to send children to the community clinic for regular growth monitoring and immunization, participate in environmental and health education campaigns, and seek immediate medical care in the event of serious health concerns. They emphasized the need for participants to see these actions as complementary to the traditional medical system, than as stand alone activities.

7.5.2 Strengthening Intersectoral Collaboration

Participants expressed the need for local government departments, the private sector, and international and non-governmental organizations working in the community to coordinate their efforts so as to increase their effectiveness of improving community well-being. Such sentiments have been expressed in the environment and health literature. For example, the ecosystem approach to human health recognizes that single-sector and disciplinary approaches to human health are ineffective to responding to the complexity of factors influencing health and environment concerns. Such collaborations do not just require commitment from participating departments, but also the active involvement of community members. In this regard, participants in the men's group suggested regular monthly meetings with all relevant institutions working in the community. They also echoed the importance of involving community members in the planning phase of community projects and providing more opportunities for community participation in decision making at the local and district level. Some women expressed the importance of including women in these meetings, as these have a high tendency of being only male-dominated. In general, participants expressed the need for interventions to be broad-based, taking into account the varying needs of all social groups and not just the so-called at-risk groups.

7.5.3 Broadening the Concept of Community Participation

Some participants pointed out that although many environment and health programs have sought to be "participatory" and inclusive in scope, they have involved community members only as providers of information and as "labourers" in the field. As one woman put it,

> participation to me means bringing us together, talking to us about the benefits of a particular project and asking us to cooperate with them.

Some participants pointed out that such narrow conceptualization of participation fails to make use of the diverse skills, knowledge base and human resource potential of community members. They made it clear that usually community efforts are beneficial when everyone has a voice, when all voices are encouraged, and when community members have the opportunity to express their views and contribute to community development in different ways.

Thus, in identifying possible strategic directions, participants expressed the need for organizations working with them to use effective participatory processes that would not only view them as sources of information, but also embrace the principles of grassroot democracy and empowerment and make use of local knowledge systems and skills. Some of the actions suggested included the formation of partnerships with extension workers, information-sharing sessions between the two parties and active involvement of community members in all phases of community projects.

7.5.4 Communicating Environment and Health Information Effectively

Participants expressed the need for the use of appropriate and effective education and communication strategies for the dissemination of environment and health information. According to some participants, health promotion programs are communicated uni-directionally, from expert to lay, with no effort being made to integrate local perspectives. Cultural dynamics, meaning and context between educators and learners, professionals and lay, play a key role in the production and acquisition of knowledge.

In commenting on the delivery of health information to lacal people, Airhihenbuwa (1994) observes that health communication projects in Africa have operated under three key assumptions: (1) health information can reach people through the media; (2) health information can change negative health practices if people have the requisite health knowledge; and (3) if relevant health information is not acquired by the people, then the skill in acquiring health messages must be developed through educational programs. The author argues that the problem with such assumptions is that, when local people are unable to acquire and practice the required health knowledge, then the responsibility for program failure is blamed on the inaction of community members. Community members are blamed for lacking the required knowledge and/or motivation to implement the necessary health actions that would lead to positive health outcomes. The health practitioner is relieved of any responsibility or accountability for the failure of participants to attain the expected health outcome. What is not examined, he argues, is the method of health communication used, and whether this is aligned with the traditional mode of learning of local communities and the cultural practices relating to how they produce and acquire knowledge. For example, in many African countries, person-to-person or home visit communication strategies have been observed to be more effective in changing negative health behaviours than mass media campaigns. Such practices are more in tune with the oral tradition as the customary mechanism of producing and acquiring knowledge in Africa (Airhihenbuwa 1994). Also educational strategies, which combine all the senses of sight, hearing, vision, taste, touch and intuition are said to be the most effective learning method (Fuglesang 1973). Many health communication programs in African communities make extensive use of posters – visual learning, which is in tune with Western culture and often requires adult literacy. In the past few decades, the use of media, especially local FM radio has been instrumental in health education, although not all rural African communities had access to electricity nor is everyone able to afford a radio.

Participants encouraged the delivery and communication of health and environmental educational programs that considered their cultural values and traditional modes of learning. Educational programs should also strengthen their understanding and knowledge base of the social and environmental determinants of health, and not focus solely on behavioural change and modification. Some women urged the use of events such as village festivals and fairs to disseminate health information. This, they suggested, could take the form of drama or specially composed songs by community groups. The men's group suggested the organization of formal award ceremonies, such as "healthy family of the month", to acknowledge families who make the effort to keep their home environments free from disease. They also suggested introducing environmental health topics earlier to school children and organizing formal debates between various community organizations.

7.5.5 Building on Local Knowledge Systems

Rajasekaran (1993) describes indigenous knowledge as a systematic body of knowledge acquired by local people through the accumulation of experiences, informal experiments, and intimate understanding of the circumstances in their culture. This collective wisdom is a rich repository of cultural norms, values and beliefs, and is acquired as people go about their daily life experiences. According to Chambers (1993), local knowledge is the basis for local-level decision making in agriculture, health care, natural resource management and a number of other endeavors in rural communities. Incorporating these varied perceptions and knowledges into programs can result in successful interventions, since the outcome often reflects the felt needs and aspirations of the people being assisted.

Despite the growing realization of the usefulness of indigenous knowledge systems, some extension workers and researchers still consider local knowledge as primitive and less valuable (Thurston 1992). Most environmental health problems are still informed by scientific understandings and western solutions. Scientific knowledge is elevated above local knowledge, and essentially fails to accommodate any local views. Participants observed that the lack of incorporation of local knowledge into community development interventions has resulted in a gradual decline of certain knowledge systems among the youth. As expressed by one woman:

> Our children are like turkeys these days. They know absolutely nothing about medicinal plants or which grass types indicate a potential good bean harvest. When I was young I knew all these plants. ...I knew what plant to use for scorpion bites. These days when a scorpion stings a child on the farm, he/she just stands there crying for help. We try to teach them about the medicinal uses of various plants but they do not listen because they think there is better medication at the clinic

Some male participants, however, attributed the lack of knowledge about some important plants among the youth to the increasing disappearance of those plant species, and suggested occasional exhibitions of important plant species to the youth. The research team followed through with this suggestion by organizing a field trip with some participants to community outskirts to collect various medicinal plants and identify their uses. Two specialists, one from the Ministry of Health and the other from the Forestry Department, joined the team to help identify the botanical names of the plants. Upon return from the trip, participants displayed and grouped the plants according to their various medicinal uses. Both the research team and the participants were amazed at the number of ailments, including malaria, diarrhoea, stomachaches, and snakebites, that could be treated with the plants collected. After taking notes on all the plant species gathered, participants were asked to take the plants home and teach their children about the medicinal uses.

During this exercise, it was interesting to see how the gendered knowledge systems on these plants played out between men and women. While most women were able to identify medicinal plants immediately at their backyards and all the way through to their bush farms, the men argued that "the proper medicinal plants were far away in the bush and not easily obtainable." Upon arrival at the farm, women readily picked up their plants, mainly leaves and branches, while the men dug deep into the soil to cut off plant roots. It was obvious that the men were very knowledgeable about the medicinal uses of roots, and the women mainly the leaves and branches. Similarly, the men demonstrated a sophisticated knowledge about "concoctions" or combinations of various medicinal plants to cure certain ailments, while the women knew about the uses of single plants. Compared to men, women showed good knowledge about plants used to cure children's ailments including diarrhoea, malaria, colds and stomachaches. Men were concerned about how to stop bleeding wounds, treat scorpion and snake bites, and how some plants could be used as pesticides and for grain preservation. In general, both men and women were very knowledgeable about these plants.

To be able to build upon this rich knowledge base of local people, participants called for fundamental changes in the way extension and community development agents worked with them. Instead of perceiving them as passive and non-knowledgeable about many events, they wanted extension workers to develop clear-cut methods for uncovering and incorporating local knowledge into intervention strategies. Participants also urged one another to continue to transmit this knowledge to future generations through oral history and folklore, and also teach their children about the importance of conserving biodiversity, as well as, the consequences of biodiversity loss for human health and well-being. This, they suggested could also be incorporated into environment and health education programs.

7.6 Building Community Capital: Integrating the Strategic Directions

Community capital is often described as comprising of four forms: *natural capital* – this comprises the natural resources that sustain our economies and health; *economic capital* – this constitutes the means by which we attain a certain level of prosperity that is sufficient to meet our basic needs; *social capital* – this deals with the social networks and systems that constitute our civic society, and *human capital* – this is reflected in our skill sets, level of education, innovativeness and creativity (Ekins et al., 1992; Hancock 2000). In most cases, communities tend to focus on increasing one form of capital, mostly economic capital, to the detriment of others. Such communities, Hancock (2000) argues, cannot be described as healthy or sustainable. In his view, a healthy and sustainable community is one that is able to increase all four forms of capital simultaneously, without hindering the development of any one capital.

In their first strategic direction, participants identified ways to improve the quality of their local environment in order to protect their ecosystems and also reduce the occurrence of diseases. In subsequent discussions, participants identified strategies meant to improve social capital. For example, they identified the need to preserve informal social networks and to create formal ones that would help community members, especially the vulnerable to meet their day-to-day needs. Participants also discussed the need to form village committees that would act as mediators between the community and the local government. These committees would then lobby the government for the extension of services such as electricity, water supply and sanitation to the community. They would also work with other government departments to influence the development of healthy public policy that would cut across a number of departments, responding to shared interests in social equity, ecosystem health and community health.

On strategies to increase economic capital, participants discussed the need to explore and implement poverty reduction strategies involving activities such as the establishment of small-scale loan programs for farmers and petty-traders, and diversification of the rural economy through the establishment of small-scale enterprises, and vocational training institutes to develop skills of local artisans, as well as explore market outlets for community products.

Strategies to improve human capital included the empowerment of women through increased enrolment of the girl-child in school, and occasional workshops by the Department of Social Welfare to enhance women's confidence and self-esteem. Some women were, however, concerned that their husbands might see such workshops as "coaching grounds" for women to rebel against their husbands, and so might prevent the women from participating. Mechanisms for ensuring social equity and conflict resolution were also suggested as ways of promoting harmony in the community, which participants saw as precursors for their community to work together to achieve a shared vision. Figure 7.2 presents the community's vision of a healthy community.

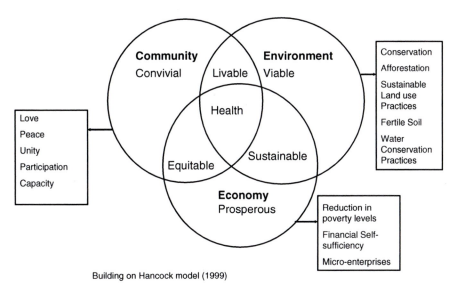

Building on Hancock model (1999)

Fig. 7.2   Envisioning a healthy community

To move these broad strategies a step closer to implementation, participants devised an implementation plan.

7.7   Drawing up an Implementation Plan

During this last phase of the strategic planning process, participants outlined, time-lines and schedules to implement the strategies and actions they identified in the previous session. They identified a series of activities to be undertaken in the coming weeks, including what will be done, who will do it, where, when, what resources will be needed, who will monitor and evaluate activities, and who will report progress to the community.

Since there were no secured funds for implementation of proposed activities, the research team encouraged participants to identify projects that could easily be implemented without any financial resources, such as the formation of lobbying groups and steering committees to lead the process. Members of the research team from the various local government departments pledged to continue to work with the community through the provision of technical assistance and guidance as they work towards achieving their vision of a healthy community.

As the principal investigator of this project and cognizant of my upcoming departure, I had mixed emotions during this last stage of the strategic planning process. I knew I could not stay long enough to see the culmination of this exercise, and so I wondered what the outcome was going to be. Would people follow through with their plans? Will they be disappointed with me coming to lead them this far and to disappear? Would subsequent issues regarding implementation generate conflict as did the other international development projects? I also reflected on my use of the participatory action research approach. Does a PAR project constitute failure, when it fails to be carried through to actual implementation? The questions were endless.

In my attempt to find answers to some of these questions, the research team held a feedback/reflection session with participants to hear their comments about the process and what they had learned. A comment from a middle-aged man summed it up:

> We did not realize we could participate in solving our own problems. For years, we have always depended on the big people in government and your people (referring to the international community) for assistance. I guess what we needed was someone to hold our hands and show us how to go about understanding and solving our own problems. This experience has been an eye-opener and we have learned a lot from you.

In general, participants seemed to have enjoyed the discussions that took place during the strategic planning process. They found the discussions interesting and informative. Some were particularly happy with the opportunity to dialogue freely with members of the research team, who until now, were seen as holding onto powerful knowledge that could not be challenged. Members of the research team seemed delighted with the process. They thought they had gained a better understanding of the community's problems and are better positioned to help them. They also learned from one another, and gained good insights into the activities of other government departments and how they might better work together to help the community achieve its dream of a healthy community. In addition, they hoped to build on the experience by continuing to work together as a team and with the community, even when the study was completed.

7.8 Conclusion

This chapter discusses the findings of a strategic planning exercise that was undertaken by community members and facilitated by the research team. The process outlined the community's vision for a healthy community, identified the barriers to achieving that vision and possible strategies to achieving it. The process also took stock of the resources and strengths of the community, outlined a plan of action, and identified which activities could be implemented immediately to begin to move the community towards their dream of a healthy community. The exercise helped participants to clearly think through how to integrate their unique strengths and to work towards becoming a better community, and better positioning themselves to confront issues of rural marginalization.

In line with the World Health Organization definition of health promotion as "a process of enabling people to increase control over and to improve their health" (World Health Organization 1986), this study, demonstrates how the involvement of local people as active participants in the assessment and identification of their own health problems not only endows them with the skills and capabilities of problem solving, but raises their consciousness about primary environmental care and sustainable ecosystem management as an important strategies to promote their health.

While this study sought to understand and intervene on the factors influencing health in one rural West African village, the findings are applicable to other communities facing similar socioeconomic, political and environmental health problems.

As have been alluded to through out the research process, fieldwork is a messy exercise, especially when dealing with an approach that requires transdisciplinarity and the use of participatory procedures. In the next chapter, I reflect on the challenges and ethical dilemmas of conducting a community-based ecosystem research.

References

Airhihenbuwa C (1994) Health promotion and the discourse on culture: implications for empowerment. Health Educ Q 21:345–353

Bergdall TD (1993) Methods for active participation. Experiences in rural development from East and Central Africa. Oxford University Press, Nairobi, Kenya

Chambers R (1993) Challenging the professions: frontiers for rural development. ITDG, London

Ekins P, Hillman M, Hutchinson R (1992) Wealth beyond measure: an Atlas of new economics. Gaia Books, London

Fuglesang A (1973) Applied communication in developing countries. The Dag Hammarskjold Foundation, Uppsala

Hancock T (1993) Health, human development and the community ecosystem: three ecological models. Health Promot Int 8:41–47

Hancock T (2000) Urban ecosystems and human health. A paper presented for the Seminar on CIID-IDRC and urban development in Latin America. Montevideo, Uruguay. April, 6–7

Kretzmann JP, McKnight JL (1993) Building communities from the inside out: a path toward finding and mobilizing a community's assets. ACTA Publications, Chicago, IL

McKnight JL (1985) Health and empowerment. Can J Public Health 76(Suppl 1):37–38

Rajasekaran B (1993) A framework for incorporating indigenous knowledge system into agricultural research and extension organizations for sustainable agricultural development in India. PhD Dissertation, Iowa State University, Ames, IA

Thurston D (1992) Sustainable practices for plant disease management in traditional farming systems. Westview Press, Boulder, CO

World Health Organization (1986). Health Promotion: Ottawa Charter. International Conference on Health Promotion, Ottawa, 17–21 November 1986. Geneva, Switzerland

World Health Organization (1994). Women's health: towards a better world. Global Commission on

Chapter 8
Challenges and Ethical Dilemmas in Conducting Participatory Ecohealth Research

Contents

8.1 Introduction

The previous chapters outlined the research methodology and processes involved in conducting an ecohealth research project, and how these were operationalized through case studies. In this chapter we step out of the research process and reflect on some of the challenges and ethical dilemmas that participatory ecohealth research projects present and how we might respond to these challenges.

There are a number of elements in an ecohealth research that present challenges, especially for individual researchers and students. First is the requirement that ecohealth research projects incorporate transdisciplinarity as a key component of the research by forming and working with a transdiciplinary research team. This requirement is very challenging for an individual researcher or a student researcher, with limited networks and institutional affiliations to execute. It is also difficult to find all the necessary experts in rural and small town settings. Although a useful requirement, transdiciplinarity is best achieved by research institutions with well-established networks and institutional affiliations. Executing it from an individual researcher perspective can be challenging and require some amount of creativity such as was carried out in the case studies described in the previous chapters.

C.Y. Dakubo, *Ecosystems and Human Health*, DOI 10.1007/978-1-4419-0206-1_8,
© Springer Science+Business Media, LLC 2011

Secondly, ecohealth research requires the use of participatory research methodologies that actively engage all stakeholders in the research process. As has been discussed in previous chapters, achieving true participation and ensuring that participatory processes do not just reflect dominant views is important, and requires the researcher to be tactful and skilled in group facilitation techniques to effectively execute this. Also participatory action research requires that action, social change, and learning be incorporated as key components of the research process. But as a researcher working within a limited timeframe and budget, to what extent should action be stated as a deliverable in a PAR ecohealth project? Does the project constitute a failure if it does not result in action? How accountable are we as researchers for ensuring that PAR projects result in meaningful learning and action, so that communities do not feel used or victimized? While the PAR research process is supposed to be collaborative and egalitarian, to what extent is the research outcome or product collectively owned? Also, how truly can we foster equal partnership between the research team and participants of the study given the varied differences, academically, socially and economically? How does a researcher deal with power dynamics between the research team and study participants?

Thirdly, as an external researcher coming into a community, how does one gain smooth entry and present your research objectives in ways that are aligned with community needs? How do you build trust with members of the research team, with the community, and with participants of the study? How will your status as an outsider interfere with the research process? Do you feel vulnerable about your limited understanding of the community's culture or about possible power struggles, conflicts and politics that could interfere with your research? How about researchers who are both insiders and outsiders, how does that interfere with the research? When do you want to be perceived as an insider, and when as an outsider? Do you even have control over how you are perceived?

In this chapter, we examine in detail some of these challenges faced by ecohealth researchers. In writing this chapter, I draw on my own research experience reflecting on some of the challenges I encountered as a doctoral candidate conducting an ecohealth research project in rural Northern Ghana, and how I responded to these. As Cotterill (1992: 602) explains, some experiences, events and feelings are unpredictable at the planning and early phases of the research and will only emerge as the fieldwork progresses. It is standing outside and reflecting on the research process that researchers are able to evaluate the research process, their relationships with the research team and study participants, and also assess the challenges and ethical dilemmas they encountered and how they responded.

8.2 Gaining a Second Entry into the Community

Selecting a site to conduct an ecohealth research project is a decision that is made based on the problem to be investigated and the availability of existing networks, collaborators and other resources. When the study site is a community, especially in a developing country, or in an Indigenous community or First Nations Reserve, gaining entry into the community and building trust is very important in carrying out a successful research project. As described in previous chapters gate keepers or existing affiliations could help facilitate entry. However, previous research affiliations that raised unexpected expectations or failed to meet the community's expectations can pose challenges. The ecohealth project I conducted in the Ghanaian village was my second encounter with the community, following an earlier project on community-based natural resource management 4 years prior.

Before starting the fieldwork, I had lots of mixed emotions. First I was happy to be returning to my native Ghana to see my family after almost 5 years of studying abroad. Second, I was excited to be going back to a community I worked with in the past and was anxious to assess the progress of the first project. Third, I wondered how I would be received by the community the second time around. I also wondered whether the community will still have the same level of enthusiasm as they did with the first project? Will the level of trust be the same or will I have to start all over building trust? I reflected on the status of my first project with them and wondered to what extent it had contributed to overall community well-being.

Upon arrival, I met with a government official who has a long working relationship with the study community. He was my first contact during the first project. The official and I followed protocol as described in earlier chapters. We met with two representatives from the community, who in turn introduced us to the village chief. Upon receiving permission from the chief to conduct the research, the gatekeepers (i.e. the two community representatives) arranged a meeting with community members to formally introduce me. This first meeting with community members was characterized by great anxiety and anticipation. First, I was anxious to see which, and how many old faces would return and be interested in working with me, since this will facilitate trust building. Also, based on events during my last visit, I thought I had developed a good rapport with them, since at the end of my stay, the research team was given a grand "send-off" party at the chief's palace, amidst singing and dancing. They even presented us with a goat as a sign of appreciation. So while I was hopeful that, this time around, residents would show their usual enthusiasm, I was also concerned about their reaction to my lack of contact with them for the past 4 years. On the one hand, I could imagine their disappointment as being victims of one of those studies. On the other hand, if my intervention generated some useful outcome or learning, then they might be receptive to working with me again.

We met with two groups – a men's group and a women's group. At the first meeting with the women's group, I was introduced by one of the gatekeepers in the following manner:

> some of you probably remember this woman; *our friend*, who came and worked with us some years back. As you remember she was a student at the time and so when she got back, *she* wrote about *her work* with us, and *her people* liked it and were very pleased with her. *She* received lots of praise for it, and so she has been asked to come and work with us *again*. I hope you will all do your best to help our friend this time around.

This introduction spoke volumes to me, and increased my apprehension and vulnerability as a researcher even more. What I understood, or at least tried to understand was that, first, the benefits of the first intervention were one-sided, accruing only to me as the researcher; second, despite all attempts to make the research participatory, egalitarian, and collaborative, it was portrayed as *her work*; and third, who did she mean by *her people*? So if the benefits of the research accrued to *a friend* who never followed-up or monitored the progress of her research to determine its impact on the community, how can we continue to help such a *friend*?

In the men's group, the old faces could barely wait for me to be introduced when they accused me of deserting them, and never bothering to "send us shirts from abroad". I used the first few minutes after each of the introductions to thank community members for coming and to explain that after my last visit I continued to study, and given the resources I had to conduct the first study, I was not able to do any follow-up studies, and so this second phase should be considered as building upon our initial experience.

This experience illustrates the challenges that can emerge when a community's expectations of a research project are not fulfilled. This community expected that I continuously remain involved in monitoring progress of my research, but given the limited timeframe and budget of a student researcher or many researchers in general, this latter phase of the project is usually not fulfilled and can result in disappointment. In moving forward with the ecohealth project, it was important to spell out the shortcomings to the community and to dispel any expectations of continuous involvement or the ability to support the implementation of any proposed activities financially. It was also important to dispel any affiliations with on-going international development projects in the community, since there is the tendency for community members to perceive all external initiatives in the community as linked one way or another, resulting in false expectations.

The second challenge was ensuring that my broader research objectives were aligned with the community's needs. From a participatory action research perspective, ideally research problems should be initiated by the community or mutually initiated by the researcher and the community. But as many PAR researchers point out, some communities usually are unaware of the problems facing them, and lack the time, resources and initial drive to begin to look for solutions to their problems. This is particularly true with many rural communities in developing countries. As such I made use of the first meeting with community members to explain the importance of understanding how we interact with our surrounding environments and how such interactions can adversely impact our health. I urged them to see me as a facilitator who was here to help them work together with other government officials to better understand the factors influencing their health and to be active participants in solving these problems. I explained that the benefits were not necessarily about the project outcome and the activities they implemented, but also about the process where community members stood to benefit from problem-solving skills, strengthened community capacity, and using the research findings to draw government support and influence policy.

8.3 Establishing Equal Partnerships Between Participants and the Research Team

Participatory action research calls for equal partnership between researchers and participants of the study, an requires that participants be seen as co-researchers and be actively involved in all phases of the research project. This requirement is theoretically feasible, but practically difficult to implement. In many settings, there are so many inherent differences between researchers and study participants that it is difficult to try to obliterate these differences and to put everyone on the same playing field. For example, researchers, by virtue of their privileged status of being educated in a particular discipline, labelled as an "expert", and possesses knowledge that is deemed legitimate, is highly ranked than the "lay" local actor. Hence, to what extent can equal partnership be achieved between the research team and community members? What role can the principal investigator play in fostering equal partnership in a research project?

In addition, members of the research team are supposed to work as equal partners and colleagues in integrating their respective knowledges to help solve the problem at hand. However, even among members of the research team, there are power dynamics due to differences in age, gender, disciplinary backgrounds, rank in institution, and relative knowledge and experience about the problem at hand. As such fostering equal partnership among members of the research team is an additional challenge, and becomes even more complicated when the principal investigator is an outsider or a student who is relatively younger and inexperienced.

In my quest to form equal partnerships with members of the research team and with study participants, I wondered how, as the principal investigator, my multiple identities as a young, female, doctoral student from Ghana studying abroad might affect this partnership, as well as the research process and outcome. Each identity changed my relationship with the study participants and members of the research team. For example, as a young student I was often referred to as our "daughter" by members of the research team. As a daughter within the Ghanaian culture, I was expected to play a submissive role and respond in ways that a daughter would normally respond to her father. Thus, from a cultural point of view, I played that submissive role, allowing myself to be deployed as such. However, this relationship changed when I had to outline the conceptual issues of the research and train members of the research team on data gathering methods such as the Strategic Planning Process. Although I was open to their opinions, I still played a lead role in specifying how the research should proceed and what to aim for. In such circumstances, equal partnership becomes elusive. Instead, we have a dynamic relationship whereby power concentrates in me at one point, to explain conceptual issues and make academic decisions, and at another point concentrates on participants indigenous knowledge systems, and on members of the research team's insights in having a close working relationship with the community. So at any one point in the research process, one group of stakeholders seem to dominate based on their privileged knowledge or expertise, further making equal partnership elusive.

There were some challenges in attempts to form equitable relationships between participants and the research team. Many members of the research team occasionally worked with community members, either by providing them with technical information, educational materials or even working with them on their farms. Most disciplinary training, especially in the medical field, instils a notion of "objectivity" in their professionals, making it difficult for them to embrace local subjectivities and give up their "expert" status (Cornwall and Jewkes 1995). The top-down fashion in which most departmental programming is implemented has impacted local people's sense of confidence in their abilities. For example, during group discussions, I noticed how some participants withheld information because they thought it might not be correct from the viewpoint of the "experts" who were facilitating these discussions. Those who responded, especially to technical questions, looked to the facilitators for approval. While participatory approaches strive to build equal partnerships and create conducive environments for open dialogue, the respective differences, and multiple and changing identities of human beings makes this difficult to achieve. Thus, the notion of equal partnership is always far from being a reality, developing sometimes during the latter stages of the research when participants become confident and feel a validation of their own knowledge systems.

8.4 Participatory Ecohealth Research as Transformative and Empowering?

In general, participatory action research aims to combine research, education and action (Hall 1992). This is based on the assumption that knowledge that is not used to stimulate social change is wasted knowledge (Israel et al. 1998; Maguire 1987). Thus, participatory research approaches aim to produce knowledge that incorporates perspectives of the marginalized, deprived, and oppressed groups of people, those experiencing a disproportionate burden of the health and environment problems, and then use that knowledge to undertake collective actions to reduce inequalities and transform social realities (Maguire 1996). Participatory research tries to give these weaker voices the opportunity to name their problems and participate in decision making towards their resolution. However, as outside researchers we are usually faced with the challenge of identifying the "silenced", "oppressed", and "marginalized". Besides, we gain entry into the community through local leaders and gatekeepers who are in a position of power and also are responsible for mobilizing community members to participate in events.

During my fieldwork, the gatekeepers I worked with, to some extent, influenced which individuals participated in the study, since they were the main contacts for the snowball sampling strategy and also responsible for announcing the study in various community groups. Hence, despite my attempts to encourage a representative sample, I felt powerless in controlling who participated, their reasons for participating and their relationships with the gatekeepers. Also, while I was interested in reaching all facets of society, especially those who encountered a disproportionate portion of the disease burden, these groups of people probably could not participate because they were either too busy securing life's basic necessities (Rifkin 1994), did not belong to any formal community organization, could not participate because of lack of clean clothes to join group discussions, or were not in good terms with those handing out the invitations. Cornwall and Jewkes (1995: 1673) observe that, "unless a definite political commitment to working with the powerless is part of the process, those who are relatively inaccessible, unorganized and fragmented can easily be left out". The challenge this situation presents is that when researchers work through local power structures, there is the risk of the research being manipulated to serve the agendas of the powerful. On the other hand, if researchers work against local power structures, there is the tendency to weaken both the potential impact of the research outcome and even intensify marginalization, alienation, and inequities after the research is completed (ibid).

What role can a researcher play to ensure that a participatory action research project results in meaningful learning and possibly action? The researcher plays the role of a facilitator, a catalyst and a co-learner. The researcher initiates the process and collectively works with the ideas of participants to achieve a new level of understanding and critical consciousness about participants' lived experiences and the broader social, economic, and political factors influencing those experiences. Such understanding is then used to influence policies or actions in ways that positively impacts the community. But as discussed previously, not all PAR projects proceed to the action stage due to limited resources, budget constraints or conflicting timeframes between the researcher and the community. The question that arises then is to what extent should a researcher, especially student researchers, specify action, monitoring and evaluation as part of the research project? Is a PAR project worthless when it fails to proceed to action?

These are some questions I took into consideration whilst forming my transdisciplinary research team. I thought it was important to include some policy makers, as well as officials from government departments with a mandate to working with communities on health, environment and natural resources issues. The involvement of such individuals in the research process gave them a better understanding of the issues facing communities, the underlying factors precipitating the problems, and the commitment of community members to finding effective solutions. This insight would likely help government departments assess how best to integrate their expertise and resources to implement the research findings. For example, diarrheal diseases, especially among children, are a major concern in the community and stem from the use of contaminated water sources and poor sanitation. One would expect ministries or government departments responsible for water supply and sanitation to work in close collaboration with the ministry of health to find effective ways of responding to such health concerns. However, such integrated working relationships is rare, primarily because many government departments work in silos and hardly incorporate health concerns into their respective mandates. The opportunity to work together on a research project enabled these departments to begin to explore how to integrate their knowledge, services, and resources in ways that would enhance community health and well-being. Such opportunities and learning go to confirm that participatory action research projects do not necessary constitute failure when they are unable to proceed to implementation.

8.5 To Be or Not To Be: Insider – Outsider Relations

Does an insider researcher or a researcher familiar with a community or issue have advantage over an outside, external researcher? What are the advantages and disadvantages of being either an insider or an outsider? Some scholars argue that "insiders", researchers who study a group, or issue to which they belong or share common identities have an advantage over "outside" researchers because "insiders" tend to draw on their knowledge of the group to gain a better understanding of informants' and participants opinions (Abu-Lughod 1988; Hill-Collins 1990). On the other hand, "outsiders" argue that by not belonging to the group under study, they are likely to be perceived as neutral, and hence be confided in on issues that would normally not be given to "insiders" (Fonow and Cook 1991). "Outsiders" also argue that they are more likely to be objective in observing people's practices and accurately interpreting them, than insiders (Mullings 1999). What both arguments assume is that the position "insider" or "outsider" is a fixed attribute that a researcher can be in or have control over.

However, according to poststructuralist theorizing, individuals possess multiple, fragmented, overlapping, and changing identities (Butler 1992). These identities, they assert, at any moment are often conflicting with one another, making it unlikely for us to be in all, or completely in any of these positions (Haraway 1988). These critical scholars point out that insider/outsider status is better perceived as "a site of betweenness" (England 1994; Kobayashi 1994; Nast 1994), or "a positional space" with unstable boundaries (Mullings 1999). As Nast (1994: 57) explains "...because we are positioned simultaneously in a number of fields we are always, at some level, somewhere, in a state of betweenness, negotiating various degrees and kinds of difference – be they based on gender, class, ethnicity, race, sexuality and so on ... as such we are never 'outsiders' or 'insiders' in any complete sense."

I find the debates on insider/outsider relationships pertinent to my experiences in the field. I reflected on how my status as a native woman from the region, living and studying abroad, might influence my working relationships with participants and members of the research team. I never perceived myself as either an insider or a complete outsider. I understood quite well that the differences between research participants and I based on my education, age, place of residence, and class outweighed our commonalities of language and place of origin. So I did not perceive myself to be in any privileged position compared to a complete "outsider". I also understood that my status as a student, and not a government official with the capacity to make financial or other resource commitment could prevent informants from confiding in me. On the other hand, my status as a Dagao[1], a clan member, and a second time researcher in the community, put me in a "somewhat" privileged position to better understand some community and cultural issues.

I chose to deploy or rather agreed to be deployed strategically either as an insider or outsider On one hand, striving to be an "insider" when this could build trust and, on the other, striving to be an "outsider" when it would alleviate my being perceived as a threat. For example, during my interviews with women, I was referred to as "our sister", "daughter" and "friend", and I strove to be those attributes based on our shared gender. By contrast, with members of the research team and community elders, I was their "daughter" who studied abroad, needed to be guided into the community, and needed assistance in completing her research. In these instances, being perceived as an "outsider" facilitated my entry into the community and accorded me the cooperation and respect normally given to "foreigners" or "visitors". My experiences illustrate that the choice to be an "outsider" or an "insider" is partly out of the control of the researcher, and may be constituted differently by different community members at different times. This further destabilizes the insider/outsider argument that assumes that the researcher has control over the choice of being either an "outsider" or an "insider".

8.6 Dealing with Vulnerability in Focus Group Discussions

In this section, I continue with my reflection by drawing attention to the fragility of the research process and how this sometimes augments the vulnerability of the researcher. I became vulnerable on a number of occasions. On one occasion, a brawl broke out during a focus group discussion with the men's group. I posed a question relating to chieftaincy issues in the community and a young man attempted to respond but was admonished by an elderly man not to "say a word" because the young participant was not born in the community, and had just returned from the "South." Besides he was too young to speak about chieftaincy matters. In response, the young man argued that, despite these shortcomings, he had the right to respond to the question since he was part of the focus group. He was further admonished by his peers for challenging an elder, and failing to respect a senior, further confirming his out of touch with the cultural values and his status as an "outsider." The exchange got out of hand and turned to a fist-fight between two young men. During the encounter, I processed a series of questions in a relatively short-time. What did I do wrong? Should I have known better because chieftaincy has always been a sensitive topic in this region? Was there an on-going chieftaincy dispute? Was there an old grudge between the two young fighters? How was I to respond? I was left powerless and wondering how to restore order to the discussion. The situation was amicably resolved by other members of the research team.

---

[1] One of the ethnic groups from the Upper West region of Ghana.

   Besides exposing my vulnerability as a researcher, this incident has other implications and lessons for participatory researchers. The first relates to the tendency to
see communities or the "field" as a homogenous, place-based entity. With respect
to this situation, factors such as where one is born and length of residency in the
community determines who belongs or does not belong to the community and what
they can and cannot say. The second relates to the structure and composition of
focus groups. Most focus groups are formed to capture diverse opinions, but from
this incident I now see focus groups as sites of power struggles among various individuals based on age, ethnicity, gender, among others. Given the cultural value not
to challenge the "received wisdom" of the elderly, how might we interpret the data
we collect from group processes? As Mosse (1994) points out, data from group
processes obscure power relations and other group dynamics and tend to reflect
the voices of those culturally sanctioned to speak in public, while silencing the
rest. These concerns challenge the view that group processes provide avenues for
the marginalized to articulate their views, be heard and build confidence to participate in decision making processes. If not careful, such processes become avenues to
redistribute, concentrate and reconfigure power in some individuals (Maguire 1996).
Finally, what is the implication for participatory processes that tend to strive for consensus and communal goals on which collective action is taken? Would such goals
simply reflect those of the powerful and the elderly? How would youth interests
come into play? Based on this then, it is important for focus groups to be formed
taking into account other issues such as age, religion, ethnicity, among others. But
even then, are there such things as "absolute commonalities"? Perhaps group processes should be deployed more strategically, weighing the magnitude and type of
the problem to be investigated, how widespread it is and the benefits associated
with mobilizing communal responses. For example, in this study, given the strong
gender relations in the community, the best way to generate good dialogue in a non-
threatening way was through gender-disaggregated focus group discussions, while
being cognizant of the diverse identities and needs of each group. The above incident also prompted follow-up, in-depth interviews with both parties of the brawl to
understand any pre-existing issues that might have triggered the confrontation.

8.7 Are all Voices and Knowledges Captured through Group Processes?

In the previous section, I alluded to the fact that the research team decided to conduct
gender-disaggregated focus group discussions in an attempt to create an environment for open and frank discussions. This is based on the assumption that each
gender group, for example, women, have more in common with each other than
they have with men and are likely to have an open and frank dialogue in a separate group discussion, and probably reach commonly shared solutions. However,
such an assumption risks smoothing over the inherent conflicts and different interests that might exist among women and men. Shared gender alone is insufficient to

organize a discussion around. As scholars informed by poststructuralist theo-rizing point out, the binary conceptualization of phenomena (e.g. men/women, lay/professional) "creates a false sense of unity by reducing the flux and heterogene-ity of experience into supposedly natural or essentialist oppositions" (Flax 1990: 36). These scholars call for different ways of researching gender-related issues by dislodging the opposition between men and women, and the assumption of simi-larities. Instead efforts should focus on showing how the complexities of specific situations of, both men and women are shaped by variables, such as ethnicity, class, religion, and place (Barnes 1982).

This false sense of commonality was revealed during focus group discussions with the women. In the women's group, socioeconomic status and material wealth played a key role in excluding people from group discussions. The women agreed to hold their group discussions on Sundays after church. In a community where most of the population is Christian, and going to church means wearing your best clothes, some women (e.g. Muslim women and traditional worshippers) did not feel comfortable participating in these discussions simply because they did not wear fashionable clothes on those days and did not want to be the odd ones out. To some extent, it is fair to assume that wearing good clothes enhances the confidence level and self-esteem of an individual, thus allowing them to dominate group discussions. In a later discussion on factors inhibiting participation in community health promo-tion programs, one female participant identified lack of clean clothes as the major obstacle stopping her from participating in such programs. In a rather subtle way, she had made an important point. Her response also confirmed my observation of how the women's group discussions had been dominated by a few individuals. The research team decided to change group discussions from Sundays after church to a different time slot in an attempt to generate good dialogue in a non-threatening environment. Although this was not a democratic move, it was observed that fol-lowing this change of day, group discussions became livelier and egalitarian, and every woman in the group laughed equally hard when there was reason to.

8.8 Conclusion

This chapter explored some of the challenges and ethical dilemmas that are some-times encountered during a participatory research project. In particular requirements such as forming a transdisiciplinary research team, ensuring equal representation in group discussions and making use of participatory procedures are fraught with challenges that require some level of creativity from researchers. These challenges were explored through my research experience and how I responded to them. I also reflected on my struggles to build trust the second time around, to strategically deploying myself as either an insider or an outsider during various stages of the research process. The chapter also raises questions about the practicality and effec-tiveness of some of the research procedures we use and whether or not they actually accomplish what they set out to. Issues of fostering equal partnership, ensuring equi-table representation, and capturing all voices, still continue to present challenges that are sometimes beyond the control of the researcher. In the next chapter we explore how the ecosystem approaches to human health might be used to examine the health issues of Indigenous communities.

References

Abu-Lughod L (1988) Fieldwork of a dutiful daughter. In: Altorki S, Fawzi El-Solh C (eds Arab women in the field: studying your own society. Syracuse University Press, Syracuse, pp 139–62

Barnes B (1982) T.S. Kuhn and social science. Macmillan, London

Butler J (1992) Contingent foundations: feminism and the question of "postmodernism". In: Butler J, Scott JW (eds) Feminists theorize the political. Routledge, New York, NY, pp 3–21

Cornwall A, Jewkes R (1995) What is participatory research?. Soc Sci Med 12:1667–1676

Cotterill P (1992) Interviewing women: issues of friendship, vulnerability, and power. Women's Stud Int Forum 15 (5/6:593–606

England KVL (1994) Getting personal: reflexivity, positionality, and feminist research. Prof Geographer 46:80–89

Fanow M, Cook J (eds) (1991) Beyond methodology: feminist scholarship as lived research. Indiana University Press, Bloomington, IN

Flax J (1990) Thinking fragments: psychoanalysis, feminism, and postmodernism in the contemporary west. University of California Press, Oxford

Hall BL (1992) From margins to center? The development and purpose of participatory research. Am Sociol 23:15–28

Haraway D (1988) Situated knowledges: the science question in feminism as a site of discourse on the privilege of partial perspective. Feminist Stud 14:575–599

Hill-Collins P (1990) Learning from the outsider within the sociological significance of black feminist thought. In: Fanow M, Cook J (eds) 1991 Beyond methodology: feminist scholarship as lived research. Indiana University Press, Bloomington, IN, pp 35–59

Israel BA, Schulz AJ, Parker EA, Becker AB (1998) Review of community-based research: assessing partnership approaches to improve public health. Annu Rev Public Health 19:173–202

Kobayashi A (1994) Coloring the field: gender, "race," and the politics of fieldwork. Prof Geogr 1:73–80

Maguire P (1987) Doing participatory research: a feminist approach. School of Education, University of Massachusetts, Anherst, MA

Maguire P (1996) Considering more feminist participatory research: What's congruency got to do with it? Qual Inq 2:106–108

Mosse D (1994) Authority, gender and knowledge: theoretical reflections on the practice of participatory rural appraisal. Dev Change 2:497–526

Mullings B (1999) Insider or outsider, both or neither: some dilemmas of interviewing in a cross-cultural setting. Geoforum 30:337–350

Nast HJ (1994) Opening remarks on "Women in the field" Prof Geographer 46:54–66

Rifkin S (1994) Participtory research and health. Proceedings of the international symposium on participtory research in Health Promotion. Liverpool School of Hygiene and Tropical Medicine. Liverpool, UK, Sept. 1993.

Chapter 9
Ecosystem Approaches to Indigenous Health

Contents

9.1 Introduction

Many Indigenous communities around the world have strong ties with the biophysical environment. As expressed in the opening chapter of this book, Indigenous communities see the forests as: "their food bank, drugstore, meat market, bakery, fruit and vegetable stand, building material centre, beverage supply, and the habitat for all of the creator's creatures."[1] These close ties with the natural environment is reflected in many aspects of the Aboriginal culture, including how health is conceptualized and experienced. Many Indigenous peoples conceptualize health from a holistic perspective and see individual and community well-being to be intricately linked to the health of the "country." Similarly, many Indigenous populations rely on

---

[1] A quote by an Aboriginal person. http://www.envirowatch.org/gndvst.htm. Accessed May 01, 2010.

C.Y. Dakubo, *Ecosystems and Human Health*, DOI 10.1007/978-1-4419-0206-1_9,   141
© Springer Science+Business Media, LLC 2011

traditional forms of healing. According to the World Health Organization (WHO), about 80% of Indigenous population in developing countries relies on traditional healing systems as the primary source of care (World Health Organization 1999).

While the natural environment has provided food, medicine and shelter for Indigenous communities, it has also become a source of disease, especially when the carefully managed ecosystem by Indigenous peoples becomes disturbed. The natural resources of Indigenous communities, including land, trees, minerals, and freshwater have become targets for exploitation by major corporations. The processes involved in logging and mineral extraction do not just destroy the ecosystem, but also the satellite factories and pulp mills established to process these raw materials discharge large quantities of toxic effluents into freshwater bodies and contaminate vital food sources. Because many Indigenous communities live on the margins of society and lack the necessary resources to effectively mitigate the adverse environmental and health impacts of these activities, they tend to be heavily impacted.

In addition, to the environmental health effects emerging from these corporate activities, many Indigenous communities, especially in rural and remote areas, continue to experience diseases of poverty. The lack of water and sanitation services predisposes Indigenous populations to high incidences of diarrhoea. Lack of adequate infrastructure and proper housing renders many Indigenous communities, especially those living in rural and remote coastal areas, vulnerable to extreme weather events and other climate change-related health impacts. Government neglect and lack of socio-economic opportunities have led to an increase in social problems such as alcoholism, smoking and high suicide rates, among many Indigenous communities, especially in North America (Bramley et al. 2004; Hunter and Harvey 2002). In addition, modern health diseases such as diabetes, cardiovascular disease, and respiratory diseases are becoming prevalent in Indigenous populations in North America and other areas in the world.

Together these environment, health, and social problems interact in complex ways to adversely impact the health status of Indigenous populations, resulting in large contrast and disparities with non-indigenous populations. For example, in Western Australia the difference in life expectancy between indigenous and non-indigenous peoples is about 20 years. Also, the complexity of the factors affecting the health status of Indigenous populations defies resolution through the use of simple, uni-dimensional, and biomedical approaches. Instead, the underlying causes of poor health outcomes in many Indigenous communities need to be examined from a political ecology of health perspective, situating these health outcomes within the context of a politicized and colonized ecosystem. Health intervention strategies must provide opportunities for Indigenous peoples to collaborate with transdisciplinary groups of professional researchers to investigate Indigenous health concerns and develop solutions that are culturally appropriate and effective. A transdisciplinary team of researchers will include Indigenous scholars who are scientifically and culturally savvy, so as to avoid a misrepresentation of Indigenous health concerns or inappropriately label or blame the "victim", while excluding the role of unequal power relations and issues of Indigenous marginalization. The integration of traditional and scientific perspectives will accommodate Indigenous conceptions of health, the intimate connections between Indigenous health and ecosystem health, as well as validate Indigenous knowledge systems. The use of participatory action research procedures will ensure that the research generates critical consciousness, generates new knowledge about the problem, and produces outcomes that are mutually acceptable both to researchers and Indigenous communities.

These features are central to the ecosystem approach to human health, which views humans as integral to nature, and seeks to promote human health and well-being through better ecosystem management. The ecosystem approach encourages integration of local knowledge systems, the use of participatory approaches, and examines how human health is shaped by the complexity of interacting factors at various spatial and temporal levels, including family, community, regional, national, and the biosphere as a whole. However, the ecosystem approach must be infused with political ecology theorizing so as to illuminate the hidden underlying forces that produce poor health in such marginalized communities. Also an incorporation of poststructuralist theorizing allows researchers to interrogate Western constructions of Indigenous peoples, their health practices, as well as the various ways in which Indigenous knowledge system are silence and Western scientific knowledge valorized.

This chapter discusses some of the environment and health challenges facing Indigenous communities and proposes a critical *ecosystem approach to Indigenous health* as an effective strategy for examining, intervening and promoting indigenous health and well-being.

9.2 Indigenous Peoples and Communities

There is an estimated 370 million indigenous peoples living in more than 70 countries worldwide. This population has over 500 languages and cultures (International Work Group for Indigenous Affairs 2001). Such diverse cultures and backgrounds makes it difficult to try to define Indigenous peoples. Any attempt to do so may reduce this diverse group into a homogenous group of people, while failing to recognize the uniqueness and differences in identity and cultures. Hence, the concept "Indigenous" has been difficult to define and remains a contested issue among many anthropologists, indigenous peoples and scholars alike (Kuper 2005). However, within the literature, the term Indigenous is used to refer to individuals officially recorded as the first human inhabitants of an area or nation prior to European colonization (Montenegro and Stephens 2006; Sylvain 2002). This distinction is particularly clear in countries such as Canada, Australia, New Zealand, the United States, and to some extent Latin America. However, in regions such as Asia, Africa, and the Middle East, the distinction between indigenous and non-indigenous peoples becomes difficult because of internal colonization and related issues, such as ethnic colonization within geographic regions, social hierarchies, apartheid, civil wars and genocides (Stephens et al. 2006).

Within the United Nations (UN) system, no official definition has been adopted, although the UN makes use of Cobo's[2] concept of "indigenous" as a working description of Indigenous peoples and communities.[3] Cobo offers the following working description for Indigenous communities, peoples and nations:

"Indigenous communities, peoples and nations are those which, having a historical continuity with pre-invasion and pre-colonial societies that developed on their territories, consider themselves distinct from other sectors of the societies now prevailing on those territories, or parts of them. They form at present non-dominant sectors of society and are determined to preserve, develop and transmit to future generations their ancestral territories, and their ethnic identity, as the basis of their continued existence as peoples, in accordance with their own cultural patterns, social institutions and legal system"[4] (United Nations 2004: 2).

An inclusive and modern understanding of "indigenous" include peoples who:

- Identify themselves and are recognized and accepted by their community as indigenous.
- Demonstrate historical continuity with pre-colonial and/or pre-settler societies.
- Have strong links to territories and surrounding natural resources.
- Have distinct social, economic or political systems.
- Maintain distinct languages, cultures and beliefs.
- Form non-dominant groups of society.
- Resolve to maintain and reproduce their ancestral environments and systems as distinctive peoples and communities (United Nations Permanent Forum on Indigenous Issues, 2004 Fifth Session, *Fact Sheet 1: Indigenous Peoples and Identity*).

9.3 Indigenous Conceptions of Health and Links to "Country"

Many Indigenous peoples conceive health from a holistic perspective (Adelson 2005; Bartlett 2005). Holistic health is concerned with addressing all facets of well-being, including emotional, social, mental, spiritual, and physical health (Walker and Irvine 1997). Health is achieved when there is balance between the body, mind,

---

[2]Jose R. Martinez Cobo was the Special Rapporteur of the Sub-Commission on Prevention of Discrimination and Protection of Minorities, and is famously known for his Study on the Problem of Discrimination against Indigenous Populations.

[3]UN Doc. E/CN.4/Sub.2/1986/7 and Add. 1–4. The conclusions and recommendations of the study, in Addendum 4, are also available as a United Nations sales publication (U.N. Sales No. E.86.XIV.3).

[4]United Nations. The concept of indigenous peoples: background paper prepared by the Secretariat of the Permanent Forum on Indigenous Issues. Document PFII/2004/WS.1/3, Department of Economic and Social Affairs, Workshop on Data Collection and Disaggregation for Indigenous Peoples, New York, 2004.

and spirit (Spector 2002). Also, health is achieved when there is harmony between individuals, their communities and the universe. Conceiving health from this perspective distinguishes it from the Western biomedical model of heath, in which the body, mind and spirit are treated as separate entities, with the body being perceived as a machine that can be fixed therapeutically.

In contrast, Indigenous conceptions of health mimic closely the broader definition of health offered by the World Health Organization's (WHO) as *a state of complete physical, mental and social well-being and not merely the absence of disease or infirmity.* Indigenous people share similar notions of health and do not conceive of health from a disease perspective. Health for many Indigenous peoples is not merely the absence of disease, but also a state of equilibrium between spiritual, communal, and ecosystem well-being (Bristow et al. 2003).

In addition, indigenous conceptions of health and well-being are closely linked with the health of ecosystems. For example, many Indigenous Australians view a connection with the "country" as a key determinant of health. Country is seen as "a place of ancestry, identity, language, livelihood and community" (Green 2008: 12.). Accordingly, if a community-owned country becomes unhealthy as a result of environmental degradation, climate change, poor environmental management, or the inability of traditional owners to continue to inhabit the land, then the people of that land will also become unhealthy (Green 2008, 2009).

Indigenous views of health are very similar to those of many African communities. Many African communities conceive of health from a holistic perspective, linking health and well-being to the health of the land, ability to fulfill family and communal obligations, ability to fulfill duties as a good wife or a responsible husband, and having access to social networks (Dakubo 2004). Such expressions of health and well-being are also congruent with findings of a study conducted with a First Nations in Saskatchewan, Canada, in which health was described in relation to physical, mental, environmental and economic needs (Graham and Leeseberg 2010).

This holistic view of health by Indigenous peoples is complemented with sophisticated understanding of traditional medicines and healing practices (Crengle 2000; Hickman and Miller 2001). Most of these medicines are derived from natural ecosystems. Many regions of the world have integrated traditional healing with the Western biomedical model of health, thus validating Indigenous knowledge systems and ways of healing. For example, in Africa, the WHO estimates that up to 80% of the African population make use of traditional forms of healing as their primary source of health care (Bristow et al. 2003).

This reverence for nature has allowed Indigenous peoples to become careful custodians of their biophysical environments, and taking care to preserve the rich biodiversity of these environments. This rich biological diversity of Indigenous ecosystems has not only been useful for traditional medicines, but also has served as the basis for most pharmaceutical discoveries (Stephens et al. 2006). However, the careful preservation of the rich biodiversity of Indigenous ecosystems has also rendered them targets for exploitation by the corporate world, including international pharmaceutical companies. The growing exploitation of both Indigenous ecosystems and knowledge systems has led to growing concerns over intellectual property and the integration of traditional and Western scientific pharmaceutical knowledge systems (Etkin 2005; Trotti 2001). The value of ecosystems and their role in preserving the health and well-being of all forms of life, including humans, led to an important declaration at the 2004 international conference on Indigenous Peoples Rights to Health which stated that: "the right to land and a healthy environment is an indispensable part of Indigenous peoples health and well-being and should be recognized" (London School of Hygiene & Tropical Medicine 2004).

9.4 Colonized and Politicized Indigenous Ecosystems: Implications for Human Health

While the dominant society has taken little interest in the well-being of Indigenous peoples, their land and the natural resources they hold are very much sought assets, with the private sector seeking to build "partnerships" with Indigenous communities to exploit those resources and promote "community development". As discussed above and reported by some Indigenous communities, many of these partnerships are not equal and the processes of resource extraction and disposal of industry waste has destroyed pristine ecosystems and caused ill health. For example, many Indigenous communities have become dumping grounds for toxic substances emanating from mining or logging activities. These chemicals have polluted water sources, contaminated food, eliminated livelihood sources, created high rates of unemployment, and caused neurological disorders (Kuhnlein and Chan 2000; Van Oostdam et al. 1999). This scenario is illustrated in one First Nations community in Northern Ontario, Canada.

9.4.1 Mercury Poisoning in Grassy Narrows First Nations

Grassy Narrows is a First Nations community in Northern Ontario, Canada. The community has a population of about 1000 people, 700 of who live on reserve. The population is comprised of 70% youth and children. The Traditional Land Use Area of Grassy Narrows is a little over 4000 square kilometers and is located in the Boreal Forest on the Precambrian Shield (Grassy Narrows Environmental Group)[5]. Grassy Narrows has many waterways and rivers flowing to the Hudson Bay. The community's forest contains medicinal plants used to treat ailments such as diabetes, heart problems, sterility, skin problems, among others. The forest also serves as a habitat for a variety of wildlife, including moose, deer, martens, eagles, rabbits, beaver, wolves, foxes, bears, and various birds. Like many Aboriginal communities,

---

[5]Grassy Narrows Environmental Group http://www.envirowatch.org/gndvst.htm Accessed April 30th, 2010.

the land has been the source of livelihood for the inhabitants of Grassy Narrows, providing food and employment through fishing.

In 1972, the people of Grassy Narrows learned that their fresh water was contaminated with mercury that had been dumped from a paper mill located 320 km upstream. Between 1962 and 1970 this paper mill released close to 20,000 pounds of mercury into the Wabigoon River. The people of Grassy Narrows, and two other communities – Wabaseemoong and Wabauskang First Nations – lived downstream and were affected by this contamination. Mercury poisoning led to the loss of commercial fishing and other traditional foods, and caused many health problems. Grassy narrows saw its unemployment levels increase from 10 to 90% with the ban of fishing. For decades, the people of Grassy Narrows have since lived with the health effects of mercury poisoning.

Recently, though, on April 7th 2010, members from the Grassy Narrows First Nations, together with supporters from the Council of Canadians, Amnesty International and several other organizations converged at the Ontario Provincial Legislature – Queen's Park in Toronto, to protest against decades of mercury poisoning. Protesters carried paper fish on sticks and about 1,000 m of blue fabric as a symbol of a "wild river". One protester summed it up this way:

> The mills take from our forest, and then give us back disease and sickness and death. Our people have suffered for 40 years from mercury poisoning, and now this sickness is being passed on to our children in the womb. We must stop the mills from destroying our forests, our water and our culture for the survival of all people (a grassroots mother and blockader from Grassy Narrows).[6]

The protest was triggered by the release of a new study on the health effects of mercury contamination on the people of Grassy Narrows, a day prior to the protest. The study was commissioned by Earthroots, an environmental group, and conducted by Dr. Masazumi Harada, a Japanese mercury expert. The findings of the study confirmed that the health effects of mercury still persisted in the community, after almost 40 years after the release of mercury into the community's waterways.

Shortly after the waterways in Grassy Narrows was contaminated, Dr. Harada, first visited the community in 1975 to investigate the cause of a number of health problems including twitches, dizziness, eye problems and severe birth defects. Through his investigation, Dr Harada determined that some inhabitants recorded mercury levels three times above those stipulated by Health Canada. In 2002, Dr. Harada returned to conduct a follow-up study. The study revealed that 43% of the people who had mercury levels above Health Canada guidelines in 1975 had died, while those inhabitants whose mercury levels were within the limits set by Health Canada were still experiencing mercury-related health problems.

Mercury is a potent neurotoxin and a persistent pollutant that can adversely impact health. Some long-term health effects of mercury poisoning, include loss

---

[6]Quote from Canadian Broadcasting Corporation News, April 2010 *Mercury Poisoning in Grassy Narrows First Nation, Canada.* Accessed April 12, 2010 http://www.cbc.ca/canada/toronto/story/2010/04/07/tor-grassy-narrows.html

of coordination, numbness in the extremities, tunnel vision, loss of balance, tremors and speech impediments. According to Dr. Harada's report, there is a high possibility for congenital minamata disease to occur in affected communities. Many of the signs and symptoms reported by the people in Grassy Narrows are consistent with minamata disease. Many pregnant women in the community reported eating fish during pregnancy and observed delayed development, seizures, cerebral palsy and other health problems in their children.

Prior to the follow-up study, in 1985, after 7 years of negotiation, Grassy Narrows accepted a compensation settlement of $8 million from the pulp mill, and the provincial and the federal governments. In 1999, Health Canada stopped systematically monitoring mercury levels in the community, based on the decision that mercury levels in the Wabigoon River had dropped below federal guidelines.

However, following the release of the recent study, which confirms the continuous presence and effects of the mercury poisoning, Grassy Narrows First Nations, together with their supporters are requesting a redress of the situation. First, they are calling on the provincial and federal governments to acknowledge the persistence of mercury-related health effects in their community. Second, they want the federal government to re-evaluate its mercury monitoring policy and tighten guidelines related to cumulative exposure to low levels of mercury (CBC News, April 7th, 2010). Third, they require that the government of Canada permanently monitor mercury levels in the community through a local environmental centre. The Chief of Grassy Narrows expressed grave concern about the situation and called on all levels of government and the medical community to take the issue seriously: "We want our issues to be dealt with seriously by the medical establishment in Canada and in Ontario."[7]

Many Aboriginal communities around the world, especially those in North America and Latin America, have lost their livelihoods to such contamination of their lands and water bodies. Extensive deforestation, clear cutting, and other land use activities have forced many communities to relocate, with some inhabiting very fragile lands that enhance their vulnerability to extreme weather events and climate change.

9.5 Climate Change and Indigenous Health

Climate change is expected to cause increases in temperature, sea level and extreme weather events. These events, in turn, will impact human health directly and indirectly, resulting in excessive heat-related illnesses, increased exposure to environmental toxins, and proliferation of vector- and waterborne diseases. Air pollutants and declining air quality could result in increases in cardiovascular and respiratory diseases (Luber and Prudent 2009; Patz et al. 2000). Climate change

---

[7]http://www.cbc.ca/canada/toronto/story/2010/04/07/tor-grassy-narrows.html   Accessed   April 10th, 2010.

will also result in climate-induced migration or displacement of people from coastal areas and island nations.

The health and physical impacts of climate change will be unevenly distributed across society, with vulnerability being determined by degree of exposure, sensitivity, and the adaptive and coping capacity of individuals and resources at communities disposal (Green 2008). In developing countries, in particular, extreme weather events such as droughts, floods and storms will augment environmental degradation and affect agricultural production and further exacerbate food insecurity. The elderly will be adversely impacted by heat waves. Many diseases, especially the well-known global killers such as malaria and diarrhoea, are very sensitive to climatic conditions, and are likely to increase with climate change.

Another group of people that will likely bear a disproportionate burden of climate change is Indigenous populations living in vulnerable areas. Indigenous people are particularly vulnerable, partly because of the disadvantaged social and economic conditions characterizing their living conditions. Many Indigenous peoples, especially those living in rural and remote communities, lack adequate housing and the necessary infrastructure to withstand the harsh effects of extreme weather events. Pre-existing psychological and physical diseases caused by poverty, marginalization and dispossession, further inhibit the ability of Indigenous peoples to adequately cope with the health impacts of climate change (Ring and Brown 2002). Some of these concerns led the Intergovernmental Panel on Climate Change's *Third Assessment Report* (TAR) to identify Indigenous peoples as one of the two most vulnerable groups to be adversely impacted by climate change (Intergovernmental Panel on Climate Change 2001). The second group is small island state populations. The World Health Organization's Commission on the Social Determinants of Health has also reiterated the need to develop interventions that will respond to the health effects of climate change on Indigenous Peoples.

Extreme weather events associated with climate change is expected to adversely impact the biophysical environment, destroying natural resources, eroding lands, and flooding landscapes. Given that many Indigenous peoples health and wellbeing are very much tied to the health of their land, the biophysical impact of climate change may result in physical, emotional and mental distress (Jackson 2005; Smith 2004). Climate change could also affect the use of the biophysical environment as a place for healing, destroy traditional food systems and disrupt certain food harvesting practices.

Cardiovascular, asthma and respiratory diseases are prevalent in many Aboriginal communities, and increase in temperature and declining air quality as a result of climate change may increase the incidence of these diseases, especially among the elderly (McMichael et al. 2003). Also communicable diseases such as bacterial diarrhoea, which is common in hot conditions, may increase with hotter temperatures (McMichael et al. 2006). Floods and storms can increase the spread of infectious enteric diseases that cause diarrhoea, especially in young children (Green 2006). Mosquito-borne diseases such as malaria, dengue, and Japanese Encephalitis tend to increase with variations in temperature, humidity, and rainfall. For example, the Torres Strait in Australia has reported increased number of deaths from malaria since 1990. The explanation of this increased incidence relates to the close location of Torres Strait to Papua New Guinea, which makes the inhabitants more vulnerable to being affected by malaria from wind-borne mosquitoes (Currie 2001). There are several other climate-sensitive diseases that could affect Indigenous peoples, including typhus, leptospirosis, and scrub (Green 2006).

Given the vulnerability of Indigenous populations to the impacts of climate change, it is important that the World Health Organization, the Permanent Forum of Indigenous Peoples, and others work with nation states to integrate Indigenous climate change concerns into national policies, putting in place the necessary measures that will prevent Indigenous peoples from being disproportionately impacted by climate change.

9.6   Examining Indigenous Health Problems from a Political Ecology of Health Perspective

The underlying causes of poor health in Indigenous communities are numerous. Some of these relate to the social and economic marginalization of Indigenous people through instruments of the colonial legacy, which further constrain access to health-enhancing resources and services. Other issues relate to the exploitation of Indigenous land and its resource base, leading to loss of livelihood, exposure to pollutants, and the emergence of new and debilitating diseases. From a discursive perspective, Indigenous peoples, their cultures, and practices have been constructed to suit the social, political, economic and academic interests of the dominant society. Some Western scientific discourses construct aspects of Indigenous practices and behaviours as "primitive", and hold them responsible for poor health outcomes. Attributing poor health outcomes or ecosystem degradation to Indigenous cultural practices and behaviours legitimizes the professional or "expert" to intervene and prescribe appropriate practices. Hence, it is through the construction of Indigenous practices and behaviours as deficient, deviating from the norm, and health-deteriorating, that intervention is justified. As discussed in previous chapters, the failure to actively involve Indigenous peoples in identifying and developing these interventions, risk developing culturally inappropriate solutions, or solutions that are unable to respond to the actual problems of Indigenous people.

Very limited research has made use of a political ecology of health framework to examine the complex interactions between Indigenous peoples and the biophysical environment and how such interactions influence Indigenous health outcomes. One of the few studies is work by Richmond and colleagues (2005). They used a political ecology of disease framework to examine First Nations perceptions of the links between environment, economy and health, and to explore the risks and benefits of salmon aquaculture for First Nations in British Columbia, Canada.

The application of a political ecology approach to health seeks to challenge the simplistic causal explanations of environmental degradation and how this differentially shapes people's health outcomes (Mayer 1996, 2000). Political ecology seeks to situate environmental problems squarely within sociopolitical, economic and historical contexts, illustrating how unequal power relations between Aboriginal actors, and the state and private sector shape people's interaction with the environment and consequently lead to environmental degradation and poor health. Political ecology of health is also critical of how Indigenous peoples are blamed for engaging in health deteriorating activities and practices, without examining the inequities and marginalization underlying these health practices.

The investigation of health problems from a political ecology perspective requires that we link the contexts in which health problems occur with the social, political and economic factors shaping people-environment interactions. Political ecology of health also requires we situate phenomena in an historical context, examining how the colonial legacy has shaped Indigenous peoples interactions with the environment, or shaped access to health care services. For example, for Indigenous peoples in North America, Australia and New Zealand, it is important to examine how the role of European appropriation of Indigenous land, the stereotyping of indigenous peoples, the extermination of whole groups, and the marginalization of Indigenous peoples from the mainstream economy all contributed to shaping the current health and environment experiences of Indigenous people. By challenging simplistic explanations of environmental degradation and poor health outcomes, political ecology situates resource extraction and environmental degradation within the context of global forces of capitalism, and corporate greed. It examines the extent to which policies, trade agreements and other forces have worked together to eliminate Indigenous entrepreneurs from actively participating in the mainstream economy, and ensuring that Indigenous communities remain at the margins of the dominant society.

The case of Grassy Narrows First Nations is a typical story that could be examined from a political ecology of health perspective. Minamata disease and other health problems facing Grassy Narrows First Nations need to be examined within the context of unequal power relations between the paper mill, Aboriginal actors and the two levels of government (provincial and federal), and how this influenced the extraction of environmental resources, and contributed to the uneven distribution of environmental costs (mercury poisoning, poor health outcomes) and benefits (corporate profits). The role of the provincial government, in granting the paper mill the rights to dump the contaminants in the Wabigoon river, as well as the federal government's decision to stop monitoring mercury levels should all be examined through a political ecology lens. Similarly, the findings of the study conducted by Earthroots should not be accepted uncritically as "fact", but also interrogated for any underlying assumptions, transparency, and an assessment of the extent to which social and political framings may have been woven into the findings.

In general, the application of a political ecology approach to environmental health allows us to critically investigate and unravel the political and hidden agendas behind the construction and explanation of environmental degradation and poor health outcomes. It goes beyond simplistic explanations of these problems, instead seeking to illustrate how these are embedded in historical, social, and political contexts. These issues are given detailed attention in subsequent chapters.

9.7 Ecosystem Approaches to Indigenous Health

A number of reports have drawn attention to the inability of conventional biomedical approaches to respond effectively to Indigenous health concerns (Stephens et al. 2006; Palafox et al. 2001). Stephens and colleagues (2006) and the United Nations Draft Programme of Action for the Second International Decade of the World's Indigenous People (2005) suggest the need for newer approaches to improving Aboriginal health. These new approaches, they argue, must provide avenues to include Indigenous peoples at all levels of the decision making process, ensure equitable access to comprehensive, community-based and culturally appropriate healthcare services, and also support the provision of health education, proper nutrition and adequate housing. In addition, they require that these new approaches Incorporate Indigenous values and knowledge systems in all policies that affect Indigenous peoples, so as to enhance their acceptability.

One emerging approach that takes into account most of these issues is the ecosystem approach to human health, also known as the ecohealth approach. The ecohealth approach is very much aligned with Indigenous perspectives of health. It emphasizes holistic notions of well-being and views human health as integral to ecosystem health (Forget and Lebel 2001). Similarly, Indigenous peoples see their health to be intricately linked to the health of the surrounding environment, and so consider it important to treat the biophysical environment with care. The ecosystem approach recognizes the interdependencies between human health and healthy ecosystems, and seeks to develop interventions that simultaneously improve the health of human beings and ecosystems (ibid). The approach considers it cost effective to promote health through better ecosystem management, than to access scarce and expensive medical services. A few studies have applied the ecosystem approach to Indigenous health issues. For example, the journal Ecohealth (2007), Vol. 4, No. 4, designated an entire issue to Indigenous perspectives.

There are three issues that are central to the ecohealth approach and are important in investigating Indigenous environmental health concerns. The first is concerned with the integration of traditional knowledge systems and scientific expertise in the investigation of environmental health problems. The ecohealth approach makes use of transdisciplinary procedures when carrying out research. Transdisciplinary processes bring together specialists from various disciplines to form a research team. This team then works with local actors and other relevant stakeholders as partners in identifying and responding to the environmental health problem at hand. In such circumstances, traditional knowledge constitutes an important component of the study and contributes to a better understanding of phenomena, such as ecosystem functioning and ecosystem changes over time. Such collaboration also increases the chance of producing culturally acceptable interventions that are likely to be implemented.

Secondly, ecohealth approaches make use of participatory procedures. Local actors are recruited as co-researchers and participate in all stages of research process, including problem identification, data gathering, data analysis and interpretation, and formulation of solutions. The use of participatory research approaches do not only result in a better understanding of the causal processes of the environmental health problem, but they also build research skills and allow local actors to own both the process and results and so are committed to implementing them.

Lastly, the ecohealth approach encourages research that is socially equitable, ensuring that the investigation of environmental health problems take into account the respective needs of all facets of the society. The approach recognizes that as a diverse population, we interact differently with the biophysical environment, thus exposing us to different health risks. For example, in many Indigenous communities, there are differences between how men and women, the youth and the elderly interact with the environment. This differential interaction endows them with different knowledges and experiences. Hence, being cognizant of these differences and multiple roles and identities and how these shape our interaction with various aspects of the environment and produce various health outcomes is important in ecohealth research.

9.8 Conducting Research with Indigenous Communities: Some Considerations

There have been many criticisms on how research is conducted with Indigenous peoples and on Indigenous issues. For the most part, criticisms have focused on the lack of opportunity for Indigenous peoples to actively participate in the research process. Instead of being active agents in investigating their own issues, they become passive "objects", involved minimally and function as information and sample providers. The use of externally-driven, top-down, and non-participatory approaches to Indigenous research stands the risk of collecting wrong information, misrepresenting and interpreting this information, and prescribing inaccurate recommendations and policy measures that fail to respond to the problems under investigation. The concern to conduct research that is meaningful and aligns with Indigenous worldviews has led to increased calls for Indigenous-led research that makes use of indigenous research frameworks (Macaulay et al. 1999; Reading and Nowgesic 2002).

In a recent study, led by a First Nations researcher, two research frameworks that are aligned with Indigenous worldviews were used to explore First Nations conceptualizations of health and well-being (Graham and Leeseberg 2010). The two frameworks included the *Kuapapa Mäori* and the *Medicine Wheel*. The *Kuapapa Mäori* is a "plan, a philosophy, and a way to proceed....strategically and purposively" (Smith 1999: 2, cited in Graham and Leeseberg 2010). This research framework has mostly been used to conduct research with the Mäori of New Zealand, and is beginning to gain widespread use in other Indigenous cultures. This research framework provides a structure for framing research questions, interacting with participants, and conducting research. The *Kuapapa Mäori* is guided by five working principles, including *Whakapapa, Te Reo, Tikanga Mäori, Rangatiratanga,* and *Whanau,* which provide guidance on conducting meaningful and ethically sound research (ibid).

The *Medicine Wheel*, on the other hand, is one of the basic symbols that depict the worldview of many Indigenous peoples, especially in North America (Svenson and Lafontaine 2003). It is often described as "an Aboriginal framework in the visual shape of a circle divided into four quadrants; each quadrant represents a direction along with the teachings for that direction" (Roberts 2005: 92). In Canada, the origin of the Medicine Wheel and how it is interpreted vary among different First Nations, although there are some themes that are commonly shared (Absolon 1993; Graham and Leeseberg 2010). For example, based on the Plains First Nations interpretation, Svenson and Lafontaine (2003: 190) identify two such themes to include: (1) everything is related to everything else, things cannot be understood outside of their context and interactions, and (2) that there are four aspects to the human condition – the physical, the emotional, the mental, and the spiritual. Graham and Leeseberg (2010) used the Medicine Wheel to organize, analyze and categorize data that was collected through *Kuapapa Mäori*. Both the *Kuapapa Mäori* and the Medicine Wheel serve as useful complements to other participatory research procedures.

The importance of conducting research that responds to the needs of Indigenous peoples, takes into account their knowledge systems and worldviews, and provides opportunities for active participation in the research process. These requirements have been expressed in guidelines throughout many institutions in the world. For example, at the national and international levels, the United Nations Permanent Forum on Indigenous Issues have outlined a series of recommendations for nation states, research groups, and non-governmental groups seeking to conduct research on Aboriginal people. One key recommendation is to ensure that research on, with, or for, Indigenous communities responds to the goals and priorities of the Indigenous communities themselves, as well as, engaging Indigenous peoples as equal partners, in all stages of data collection, including planning, implementing, analyzing and dissemination (United Nations 2004).

In Canada and other parts of the world, various institutions, including the Indigenous Peoples' Health Research (Ermine et al. 2004), the Canadian Aboriginal Health Organization (Schnarch 2004), and the Canadian Institutes of Health Research (CIHR), have all developed research guidelines and ethical principles for conducting Indigenous-related research.

Some of the key issues to take into account are discussed below.

(1) *The Right to Participate*: In the past, and to a lesser extent currently, Indigenous peoples have been objects of study by outside "experts." These approaches rarely provide opportunities for Indigenous peoples to become active agents in search for solutions to their own problems. It is important that an "active offer" be made to Indigenous peoples to participate in any research endeavour that petain to them or their territory.

(2) *Engagement as Co-Researchers*: Participatory research approaches are those that provide the opportunity for those whose issues are being investigated to partipate as co-researchers and equal partners in the research process. Community members identified to participate in the study should be involved in all stages of the research process; from identification of the problem, through data collection, analysis, and implemetation of the findings.

(3) *Equal Partnerships*: True collaborative research ensures that there is equal partnership between community members and other researchers. Such equal partnership is achieved: where power and leadership for the research process is shared, where there is mutual respect for one another's views, cultures and practices; and where the research and its outcome are of mutual interest and benefits to both communities and researchers. Should community members elect not to be equal partners in certain aspects of the research, especially in technical issues, this option should be respected. However, it is prudent that researchers ensure that community members are well-informed about that aspect of the research, so as to be able to make an informed decision regarding the research outcome.

(4) *Community Consultation*: Indigenous communities must consent to any research that has to be undertaken in their community or about them by outsiders. Identifying how to gain entry into the community, identifying who to consult with, engaging in meaningful consultation, and gaining trust and building relationships and partnerships are all essential requirements to conducting a meaningful research.

(5) *Sharing Research Benefits*: It is important to determine from the outset whether the proposed research will benefit the community. Benefits from a research project may take various forms, including tangible and non-tangible benefits, immediate and long-term, monetary and non-monetary, and shared access to the findings of the study. Whatever form the benfits take, it is important to undertand how communities are intepreting this benefit and how this will contribute to community well-being. To the extent that a report, educational materials or other relevant documents are produced from this study, the researcher(s) must make the effort to translate these into the language of the Indigenous community and make them available in places that many community members can benefit from.

(6) *Capacity Building*: The objective of actively engaging Indigenous people in all aspects of the research is to create awareness about the problem under investigation and build research skills in community members. Research projects that make use of data gathering procedures like workshops, search conferences and focus group discussions must make the effort to train Indigenous co-researchers how to use these tools, so that they can use them to investigate similar problems in the future.

(7) *Acknowledging Indigenous Knowledges as Valid Ways of Knowing*: A truly collaborative research is one that respects and acknowledges participants' views as valid ways of knowing. The failure to agree on an issue does not mean one's knowledge is inferior and the other superior. It simply means we have different ways of viewing and interpreting reality, and acknowledging that reality as socially constructed is an important aspect of egalitarian research.

(8) *Issues of Intellectual Property*: Indigenous people have valuable knowledges, resources, and artifacts that have served society in valuable ways, including many drug and pharmaceutical discoveries. Many Indigenous communities also harbour sacred sites that are well preserved. It is important to recognize that Indigenous peoples and their communities have the inherent right to all these

assets, which they may choose to share or not share with external researchers. It is important that researchers respect these rights, and agree prior to the research to not take away or disclose protected and coveted information about any of these assets without the community's consent.

(9) *Cultural Competency*: Indigenous communities have a rich culture. Indigenous practices, views, and social norms are quite distinct from the non-Indigenous population, and external researchers have the obligation to educate themselves about Indigenous cultures, so as to be able to conduct culturally respectful and useful research, thereby building long-term relationships with the community and being accepted as an extended community member long after the research is over.

9.9 Conclusion

This chapter illustrates that the underlying causes of poor health in Indigenous communities are complex and defy conventional biomedical interventions. Newer approaches that integrate a variety of perspectives, including the ecosystem approache to human health, a political ecology of health, and community-based participatory processes will help unravel and respond to the complex forces inter-acting to adversely impact the health status of Indigenous peoples. The ecosystem approach emphasizes transdisciplinary procedures that integrate local knowledge systems with scientific knowledge from the natural, social, and health disciplines to allow for a comprehensive understanding of phenomena. The emphasis on partici-patory approaches and sensitivity to social equity allows for research that responds to Indigenous needs and provide opportunities for active involvement. However, the ecohealth approach alone is insufficient to unravel the unequal power relations that work together with other forces to produce environmental degradation, poor health outcomes, and the construction of Indigenous people and their communities and environments as deficient and in need of repair. The discourses surrounding Indigenous environment and health issues must be interrogated for their social and political framings, and the mechanisms through which Indigenous peoples are produced and reproduced. Understanding how these processes play out pro-vides an entry point to re-engage Indigenous peoples as active players in their own development. The ecohealth approach can be a powerful analytical frame-work, when infused with poststructuralist political ecology and critical public health perspectives.

References

Absolon K (1993) Healing as practice: teachings from the Medicine Wheel. A commissioned paper for the WUNSKA network, Canadian Schools of Social Work. Unpublished manuscript
Adelson N (2005) The embodiment of inequity: health disparities in aboriginal Canada. Can J Public Health 96:S45–S59

Bartlett J (2005) Health and well-being for Metis women in Manitoba. Can J Public Health 96: S22–S27

Bramley D, Hebert P, Jackson R, Chassin M (2004) Indigenous disparities in disease-specific mortality, a cross-country comparison: New Zealand, Australia, Canada, and the United States. N Z Med J 117:U1215

Bristow F, Stephens C, Nettleton C, W'achil U (2003) Health and wellbeing among Indigenous peoples. Health Unlimited/London School of Hygiene and Tropical Medicine, London

Crengle S (2000) The development of Maori primary care services. Pac Health Dialogue 7:48–53

Currie B (2001) Environmental change, global warming and infectious diseases in Northern Australia. Environ Health 1:34–43

Dakubo C (2004) Ecosystem approach to community health planning in Ghana. EcoHealth 1: 50–59

Ecohealth Journal (2007) Indigenous Perspective. Springer, New York 4(4):369–536

Ermine W, Sinclair R, Jeffrey B (2004) The ethics of research involving Indigenous peoples. http://www.iphrc.ca/text/Ethics%20Review%20IPHRC.pdf. Accessed 12 May 2010

Etkin NL, Elisabetsky E (2005) Seeking a transdisciplinary and culturally germane science: the future of ethnopharmacology. J Ethnopharmacol 100:23–26

Forget G, Lebel J (2001) An ecosystem approach to human health. Int J Occup Environ Health 7(2 Suppl):S3–S38

Graham H, Leeseberg L (2010) Contemporary perceptions of health from an indigenous (Plains Cree) perspective. J Aboriginal Health 6(1):6–17

Green D (2006). Climate change and health: impacts on remote Indigenous communities in northern Australia. Melbourne: CSIRO, 2006. CSIRO and Australian Bureau of Meteorology. Climate change in Australia. http://climatechangeinaustralia.gov.au. Accessed May 2010.

Green D (2008). Climate impacts on the health of remote northern Australian Indigenous communities. In: Garnaut climate change review. Canberra: Australian Government Department of Climate Change. http://www.garnautreview.org.au/CA25734E0016A131/WebObj/03-CIndigenous/$File/03-C%20Indigenous.pdf. Accessed 10 May 2010.

Hickman MS, Miller D (2001) Indigenous ways of healing guinea wormby the Sonninke culture in Mauritania, West Africa. Hawaii Med J 60:95–98

Hunter E, Harvey D (2002) Indigenous suicide in Australia, New Zealand, Canada, and the United States. Emerg Med (Fremantle) 14:14–23

Intergovernmental panel on climate change (2001) Climate change 2001: impacts, adaptation and vulnerability. Contribution of working group II to the third Assessment Report of the IPCC. In: McCarthy J, Canziani O, Leary N, Dokken D, White K (eds) Cambridge University Press, Cambridge, UK). www.ipcc.ch/pub/reports.htm. Accessed 2 May 2010.

International Work Group for Indigenous Affairs (2001) The Indigenous world 2000/2001. International Work Group for Indigenous Affairs, Copenhagen

Jackson S (2005) A burgeoning role for aboriginal knowledge'. ECOS June–July 2005:11–12

Kuhnlein HV, Chan HM (2000) Environment and contaminants in traditional food systems of northern indigenous peoples. Annu Rev Nutr 20:595–626

Kuper A (2005) Indigenous people: an unhealthy category. Lancet 366:983

London School of Hygiene & Tropical Medicine, Health Unlimited (2004). Indigenous peoples' right to health conference and public meeting. London

Luber G, Prudent N (2009) Climate change and human health. Trans Am Climatol Asso 120: 113–117

Macaulay AC, Commanda LE, Freeman WL et al (1999) Participatory research maximises community and lay involvement. North American Primary Care Research Group. BMJ 319: 774–778

Mayer JD (1996) The political ecology of disease as one new focus for medical geography. Prog Human Geogr 20(4):441–456

Mayer JD (2000) Geography, ecology and emerging infectious diseases. Soc Sci Med 50(7–8):937–952

McMichael A, Woodruff R, Whetton P, Hennessy K, Nicholls N, Hales S, Woodward A, Kjellstrom T (2003) Human health and climate change in Oceania: a risk assessment 2002. Department of Health and Ageing, Canberra

McMichael A, Woodruff R, Hales S (2006) Climate change and human health: present and future risks. Lancet 367:859–869

Montenegro R, Stephens C (2006) Indigenous health in Latin America and the Caribbean. Lancet 367:1859–1869

Palafox NA, Buenconsejo-Lum L, Ka'ano'I M, Yamada S (2001) Cultural competence: a proposal for physicians reaching out to Native Hawaiian patients. Pac Health Dialog 8:388–392

Patz JA, Engelberg D, Last J (2000) The effects of changing weather on public health. Annu Rev Public Health 21:271–307

Reading J, Nowgesic E (2002) Improving the health of future generations: the Canadian Institutes of Health Research Institute of Aboriginal Peoples' Health. Am J Public Health 92:1396–1400

Richmond C, Elliott SJ, Mathews R, Elliott B (2005) The political ecology of health: perceptions of environment, economy, health and well-being among 'Namgis First Nation's. Health Place 11(4):349–365

Ring I, Brown N (2002) Indigenous health: chronically inadequate responses to damning statistic. Med J Aust 177:629–631

Roberts R (2005). Stories about cancer from the Woodland Cree of northern Saskatchewan. Unpublished Doctoral dissertation, University of Saskatchewan

Schnarch B (2004) Ownership, control, access, and possession (OCAP) or self-determination applied to research: a critical analysis of contemporary First Nations research and some options for First Nation communities. National Aboriginal Health Organization. Ottawa, ON

Smith B (2004) Some natural resource management issues for indigenous people in Northern Australia. Paper Presentation. Arafura Timor Research Facility Forum, ANU

Smith LT (1999). Kaupapa Maori methodology: our power to define ourselves. A seminar presentation to the school of education, University of British Columbia

Spector RE (2002) Cultural diversity in health and illness. J Transcult Nurs 13(3):197–199. Retrieved September 3, 2006 from CINAHL database

Stephens C, Porter J, Nettleton C, Willis R (2006) Disappearing, displaced, and undervalued: a call to action for Indigenous health worldwide. Lancet 367:2019–2028

Svenson KA, Lafontaine C (2003). Chapter six: the search for wellness. In First Nations and Inuit Regional Health Survey Steering Committee, First Nations and Inuit Regional Health Survey. First Nations and Inuit Regional Health Survey Steering Committee, Ottawa, ON

Sylvain R (2002) Land, Water and Truth. Am Anthropol 104:1074–1085

Trotti JL (2001) Compensation versus colonization: a common heritage approach to the use of indigenous medicine in developing Western pharmaceuticals. Food Drug Law J 56:367–383

United Nations (2004). The concept of indigenous peoples: background paper prepared by the Secretariat of the Permanent Forum on Indigenous Issues. Document PFII/2004/WS.1/3, Department of Economic and Social Affairs, Workshop on Data Collection and Disaggregation for Indigenous Peoples, New York, NY

United Nations (2005). United Nations General Assembly. Draft programme of action for the Second International Decade of the World's Indigenous People. Report of the Secretary General. Indigenous Issues: United Nations 2005:7

United Nations Permanent Forum on Indigenous Issues (2004) Fifth session, fact sheet 1: indigenous peoples and identity

Van Oostdam J, Gilman A, Dewailly E et al (1999) Human health implications of environmental contaminants in Arctic Canada: a review. Sci Total Environ 230(1):82

Walker D, Irvine N (1997) Lokomaika`I (Inner Health) in a remarkable hospital. Nurs Manage 28(6):33–36

World Health Organization (1999). The Health of Indigenous Peoples – WHO/SDE/HSD/99.1

Chapter 10
Policy Frameworks on Health and Environment Linkages

Contents

10.1 Introduction

The linkages between human health and environmental conditions are well established. Yet the extent to which environmental actions and health actions are coordinated to jointly respond to environment and health challenges is still limited. This limited capacity to develop integrated environment and health policy frameworks is particularly eminent in developing countries, where many sectors still prefer to develop policies in silos despite the apparent linkages or implications with other sectors. In addition, this challenge is augmented by the sectoral institutionalization and prioritization of health and environment in different regions of the world. For example in Africa, the health sector is highly prioritized over the environment sector (WHO Regional Office for Africa 2009), despite the fact that many of the major killers, including malaria, diarrhoea, and respiratory infections are strongly influenced by environmental factors. Similarly, diseases such as diarrhea, cholera and other water-borne diseases can be reduced drastically through the provision of safe drinking water, adequate water supply, and good sanitation. The responsibility for providing such basic services often fall outside the core mandate of the health sector, residing with other departments whose policies, albeit with health

C.Y. Dakubo, *Ecosystems and Human Health*, DOI 10.1007/978-1-4419-0206-1_10,   159
© Springer Science+Business Media, LLC 2011

implications may be developed with minimal consultation or input from the health sector. Also, land use activities such as dam construction, irrigation, mining and logging have major public health implications, yet the policies surrounding these activities give little attention to the public health impacts. As our understanding of the interdependencies between health and environment increases, it is important that we find effective ways of developing integrated policy frameworks, that will take into account the activities of related sectors.

For the past two decades, a number of initiatives at the international, regional, and national levels have drawn attention to the importance of developing integrated policy frameworks on health and environment, and using these to inform other public policies to promote healthy living. Below we examine some of these initiatives, including the issues and challenges pertaining to the development and implementation of such integrated policy frameworks.

10.2 Global Policy Frameworks on Health, Environment and Development

At the global level, a number of initiatives have drawn attention to the intricate linkages among health, environment and development. These initiatives have since laid the foundation for subsequent regional and national efforts. Although a number of initiatives have been in place prior to the 1992 Earth Summit in Rio de Janeiro, it was the Earth summit that introduced the concept of sustainable development and emphasized the connections between social, economic, and environmental components of development. These linkages were expressed through the first principle of the Rio Declaration on Environment and Development, which stated that: *Human beings are at the centre of concerns for sustainable development. They are entitled to a healthy and productive life in harmony with nature.* This principle underscores the interdependence of human health and ecosystem health, and points to how all dimensions of development centres around human health and well-being, as discussed in Chapter 2 of this book.

The Action Plan for the Earth Summit, Agenda 21 further elaborated on these linkages. In particular, Chapter 6 of Agenda 21 which focuses on the protection and promotion of human health, places emphasis on the following program areas: meeting primary health care needs, particularly in rural areas, controlling communicable diseases, protecting vulnerable groups, meeting the urban health challenge, and reducing health risks from environmental pollution and hazards.

The United Nations Millennium Development Goals (MDGs), adopted in 2000 by member states of the UN, emphasize the linkages between health and environmental conditions, and view integrated policies as essential to achieving the MDG targets by 2015. In particular, three of the MDGs are related to health, and include *reducing child mortality*; *improving maternal health; and combating HIV/AIDS, malaria and other diseases*; while one MDG relates to the environment, and include *ensuring environmental sustainability.* However, given that environmental factors are responsible for up to 25% of the global burden of disease, the health-related MDGs cannot be achieved without improved environmental conditions and other socio-economic factors. In order to promote sustainable development and human well-being, both environment and health MDGs must be nurtured alongside other MDGS and incorporated into broader poverty reduction strategies.

For example, MDG 4 - *reducing child mortality*, aims to reduce by two-thirds between 1990 and 2015, the under-five mortality rate. However, close to 60% of infant mortality in developing countries is linked to infections and parasitic diseases, such as diarrhoea, cholera, malaria and acute respiratory infections (ARIs), which tend to be influenced by poor environmental conditions and exacerbated by poverty. According to the World Health Organization, about 50% of deaths from lower respiratory infections are preventable by eliminating indoor air pollution caused by solid fuels. Also, providing safe drinking water, sanitation, and hygiene can prevent about 88% of deaths from diarrhoea (WHO 2006). Thus, a coordinated approach to responding to environment and health problems is an effective strategy to meeting the related MDG targets and goals.

Similarly, MDG 5 is concerned about *improving maternal health*, and has a set target to reduce by three-quarters, the maternal mortality rate, between 1990 and 2015. In Africa, women face a 1-in-13-lifetime risk of dying during pregnancy and childbirth. Nutritional deficiencies and occupational health risks pose serious threats to both the unborn child and mother during pregnancy. Indoor air pollution from biomass burning increases the vulnerability of both women and children. Effective prevention and the attainment of other MDGs such as achieving universal primary education, especially for girls; and the promotion of gender equality and empowerment of women could contribute in reducing maternal mortality.

MDG 6 is concerned with *combating HIV/AIDS, malaria and other diseases*, and by 2015 hopes to stop and begin to reverse the spread of these diseases. HIV/AIDS and malaria are among the biggest killers in the developing world. In 2008, malaria alone accounted for 247 million cases with about 1 million deaths, affecting mostly children in Africa. According to the WHO, a child in Africa dies from malaria every 45 seconds (WHO Factsheet on Malaria 2010). However, while mosquito nets and drugs seem to be the favoured therapy for malaria, there is growing realization that the malaria can be reduced substantially through environmental enhancements that eliminate mosquito breeding grounds as well as improvements to overall living conditions.

Finally, MDG 7 is concerned with *environmental sustainability*, and aims to integrate the principles of sustainable development into country policies and programs, and also reverse the loss of environmental resources. As has been discussed throughout this book improved human health is inextricably linked with healthy ecosystems. The preservation of biodiversity is important for the discovery of new therapies for diseases such as HIV/AIDS, malaria and other ailments. The biophysical environment continue to be the major source of medicinal plants and healing for many Indigenous communities and the preservation of environmental resources, together with the provision of other basic services will go a long way to improving human health. Another target of MDG 7 is to reduce by half the proportion of people without sustainable access to safe drinking water and basic sanitation by 2015. Current estimates indicate that about one billion people lack access to safe drinking water and almost 2 billion lack adequate sanitation. These conditions have enormous implications for diseases such as diarrhoea, which can substantially be reduced through the provision of safe drinking water.

The intricate linkages among the various MDGs illustrate the importance for intersectoral collaboration, and the development of integrated policies to promote sustainable development. In addition, given the important role of environmental conditions in achieving the health-related goals, there is the need to ensure that national programs adopt integrated environment and health policies.

Another global initiative that proposed a joint approach to examining the essential elements for health promotion and sustainable development is the Johannesburg Plan of Implementation (JPI). The JPI was adopted at the World Summit on Sustainable Development in Johannesburg (WSSD 2002). Following on the footsteps of the Rio Summit, the World Summit on Sustainable Development in Johannesburg re-iterated the linkages between human health and environmental conditions and stressed the need for concerted efforts to integrate health and environment actions. The Johannesburg Plan of Implementation urges actions to address the causes of ill health, including environmental causes and their impact on development, and paying particular attention to women and children. The plan called on countries to use chemicals in ways that will have minimal effect on human health. Chapter 5 of the plan, which is on health and sustainable development, called for the reduction of respiratory diseases and other health impacts resulting from air pollution, especially in women and children. The Plan also calls for the need to strengthen the capacity of health care systems to deliver basic health services to all facts of the population in an efficient, accessible and affordable manner (Chapter 6, Sect. 54). In Chapter 8 of the plan, which is on Africa and sustainable development, the international community committed to helping and strengthening health systems that promote equitable access to health care services, as well as promote indigenous knowledge, and build the capacity of medical personnel.

In addition to formal agreements, a number of initiatives were launched at the margins of the WSSD summit, including the Healthy Environments for Children Alliance that is aimed at addressing priority health risks of children. Also, as part of the Johannesburg summit, an initiative, known as the *WEHAB Initiative* was proposed by the then UN Secretary General, Kofi Annan. The primary objective of the *WEHAB Initiative* is to draw attention and seek action on issues that are deemed essential and integral to a coherent approach to implementing sustainable development. These issues have been identified as part of the Johannesburg Plan of Implementation and focus on five areas: Water, Environment, Health, Agriculture, and Biodiversity and Ecosystem management (WEHAB). The initiative is also integral to the implementation of the objectives of Agenda 21.

For each theme, a framework of action has been developed. These thematic frameworks draw attention to the emerging issues and challenges facing each sector and identify action-oriented strategies. For example, the theme on Water draws attention to water and sanitation-related health conditions and identifies strategies for addressing these concerns. For the health theme, a *Framework for Action on Health and Environment* emphasizes the intricate linkages between health and environment, identifying action for areas that include: diseases and conditions resulting from degraded environments, occupational health and safety, health issues facing vulnerable populations, and attention to women's health concerns. The strategies identified to resolve these health conditions included among others, the need for political will and long-term commitments, action on health and environment linkages, sound policies and strategies based on the best scientific evidence, clear targets and time frames for monitoring and evaluation, and capacity-building in countries. Other strategies included: advancing research and development, mobilizing financial and human resources, and exploring intersectoral action and partnerships among key stakeholders (WEHAB Working Group on Health 2002). Overall, the WEHAB Initiative drew attention to the need for a concerted approach to fostering development by actively responding to the challenges facing the vital sectors responsible for human and ecosystem health.

Since the Summit in Rio, there have been many more recent initiatives calling for increased integration of health and environment issues.

10.3 Regional Initiatives on Health and Environment: Africa and Europe

To complement international initiatives on integrated approaches to health and environment, a number of regional initiatives have emerged in the past few decades, especially in regions such as Latin America, the European Union (EU), and Africa. Some of these initiatives have taken the form of inter-ministerial (health and environment ministers) meetings that bring together country delegates from within the region to explore commonly shared environment and health challenges and to find integrated solutions to them. The level of capacity for integrated policy development varies from region to region. Compared to Africa, Europe has had a long history of developing and implementing integrated approaches to environment and health issues. For example, while Africa held its first inter-ministerial conference on health and environment in 2008, Europe held its fifth in March 2010 in Parma, Italy, with the first in 1989 in Frankfurt. In addition, the EU has developed strategies and action plans such as the *European Union Environment and Health Strategy (2003)* and the *European Union Environment and Health Action Plan 2004–2010*. The European Union can be said to have paved the way for regional initiatives on health and environment. It is also using its expertise to help other regions respond to regional environment and health challenges. For example, through the Africa-Europe Strategy and Partnerships, the EU and Africa are joining efforts to address environmental health challenges through support with the implementation of water and sanitation programs and projects. With emerging new environment and health challenges, it is expected that many regional initiatives will emerge to explore effective and integrated approaches to investigating and addressing responding to these new environmental threats. Below we examine some key environment and health initiatives being implemented by Africa and the European Union.

10.3.1 Health and Environment Policy Frameworks in Africa

Over 23% of deaths in Africa, estimated at about 2.4 million a year, are attributed to environmental risk factors, affecting mainly the poor and vulnerable (WHO 2006). While this environmental burden of disease in Africa is well-acknowledged, there seem to be limited capacity to respond to environmental health challenges in a coordinated way. This challenge has been attributed partly to the separate institutionalization and prioritization of the health and environment sectors within African leadership structures. For example, the African Union (AU), with a membership of 53 countries, is the principal organization responsible for the promotion of the socio-economic development across the continent. The AU has a Conference of Ministers of Health (CAMH) and the African Ministerial Conference on the Environment (AMCEN). For several years, both bodies have sought to promote environment and health priorities in their respective sectors. However, it was in August 2008, when the first Inter-ministerial Conference on Health and Environment in Africa was held in Gabon. The conference, which brought together 22 ministers of environment and 26 ministers of health, adopted and signed the Libreville Declaration on Health and Environment in Africa. Nine thematic papers on health and environment were discussed, including issues related to: climate change, new and emerging environmental threats to human health, traditional and current environmental risks to human health, the economic and development dimension of environmental risk factors to human health, and the contribution of ecosystem services to human health and well-being. Others included health impact assessment, tools and approaches for policy making in environmental risk factors, and international legislative and regulatory tools for addressing health and environment challenges (International Institute of Sustainable Development (IISD) 2008). In addition to the above thematic areas, there were also side events on Children's health and environment, and a launch of a joint WHO-UNEP Health and Environment Linkages Initiatives Toolkit.

The Libreville Declaration saw African ministers commit their countries to establishing a health-and-environment strategic alliance, which would serve as the basis for national plans for joint action. Ministers requested assistance from the World Health Organization and the United Nations Environment Programme (UNEP), alongside other partners, to support the implementation of the 11 priorities identified in the declaration, and to build the capacity of African countries to conduct applied research, and to track and monitor environmental determinants of health in the region. A proposal to develop an African network for surveillance of communicable and non-communicable diseases, especially those with environmental determinants was suggested.

In response UNEP and WHO have provided guidance for the development of situation analyses and needs assessment to assist with the preparation of national plans of joint actions to implement the 11 priorities. In addition, a data management system for health and environment linkages has been developed to ensure that data collected from various countries are standardized. African countries are also being guided on the development and implementation of national plans of joint action. These plans will be implemented based on country situational analyses and needs assessment reports to ensure a comprehensive integration of health and environment into development policies and plans.

Health and Environment also are among the priority areas of the New Partnership for Africa's Development (NEPAD). NEPAD is a program that was adopted by the African Union in 2001 to respond to the development objectives of the AU. The objective of NEPAD is to promote and lead the development process in priority areas such as health, education, information and communication technology, among others. The goal is to become more proactive in developing strategies to eradicate poverty and place Africa on a path to sustainable growth and development.

Health and Environment objectives are articulated in two NEPAD documents: the *NEPAD Health Strategy* and the *Environment Action Plan of NEPAD*. The Health Strategy recognizes the broader socio-economic and political factors that underlie much ill health in Africa and proposes a multi-sectoral approach to addressing the region's disease burden. The strategic directions outlined in the document include the following:

- Strengthen commitment and stewardship roles of governments, and harness a multi-sectoral effort;
- Strengthen health systems and build evidence-based public health practice;
- Scale up communicable and non-communicable disease control programs, especially recognizing the unprecedented challenge posed by HIV/AIDS, tuberculosis and malaria;
- Reduce conditions associated with pregnancy and childbirth;
- Empower individuals, families and communities to act to improve their health;
- Share available health services equitably within countries; and
- Mobilize and effectively use sufficient sustainable resources.

NEPAD proposes that health interventions deal with the underlying determinants of health, including poor governance, socio-political instability, economic underdevelopment, poverty, marginalization, lack of infrastructure, low educational levels, low agricultural productivity, environmental degradation and gender and other social inequalities. This is based on the recognition that disease-focused interventions alone are unable to achieve sustainable human and health development.

The *Environment Initiative* identifies the root causes of most environmental degradation to include the complex interplay between poverty and excessive use of the natural resource base. The initiative proposes a coherent action plan and strategy to address the region's environmental challenges while simultaneously combating poverty and promoting socio-economic development.[1] Among the issues to be addressed are:

- Combating land degradation, drought and desertification;
- Conserving Africa's wetlands;
- Preventing, control and management of invasive alien species;
- Conservation and sustainable use of marine, coastal and freshwater resources;
- Combating climate change in Africa;
- Cross-border conservation or management of natural resources;
- Cross-cutting Issues;
  - Poverty and Environment
  - Environment and Health
  - Transfer of Technology

NEPAD's Environment Action Plan explicitly acknowledges the impact of the environment on health and points to the linkages between biophysical and anthropogenic factors, and their influence on human health and well-being. The Environment Action Plan identifies priority areas related to chemical contamination and management of pollution, including pollution of agrochemicals; and industrial, coastal, and freshwater pollution. The plan emphasizes the impact of climate change on vector and water-borne diseases and encourages the integration of health and environment policies.

Both the Health Strategy and the Environment Initiative acknowledge the intricate links among health, environment and development, and the need to adopt an integrated and multi-sectoral approach to addressing problems at the interface of health and environment. However, like many other initiatives, NEPAD faces a challenge of translating this understanding into integrated policies and programs that will respond to regional environmental health problems in a coordinated manner. For example, in the health strategy, despite the acknowledgement of malaria, HIV/AIDS and tuberculosis as key health concerns, most of the proposed interventions fall predominantly within the health sector and focus heavily on improvement of the health system. Similarly, the projects identified under the *Health and Environment* sub-theme of the Environment Initiative focus excessively on chemical contamination and management, while paying little attention to other environmental factors. These challenges are further augmented by the separate institutionalization of both the Environment and Health strategies, as well as the lack of research capacity to gather evidence to inform the development of integrated health and environment policies.

The World Health Organization (WHO) Regional Committee for Africa is the WHO's Governing Body for Africa with the mandate to develop regional health policies and programs. In 2002, the Committee adopted an environment and health strategy called *Resolution on Health and Environment: A Strategy for the African*

---

[1]UNEP/NEPAD 2003: Action plan of the Environment Initiative of the New Partnership for Africa's Development (NEPAD)

Region. The primary objective of the resolution is to enhance the capacity of African countries to improve the health status of Africans through the development and implementation of policies and advocacy in managing environmental health concerns. In addition, the WHO Regional Office for Africa's Environment and Health Strategy aims to create, by 2020, an enabling environment that promotes health and contributes to sustainable development in the continent. Among other things, by 2010, the strategy aims to help countries develop their own policies on environmental health, establish and strengthen appropriate structures for environmental health services, foster sector collaboration and partnerships, and improve human resource capacities in environmental health.

Following these policy initiatives, a number of African countries have embarked on programs that aim to integrate health and environment concerns into national development programs, and link these with ongoing projects related to the Millennium Development Goals, the Johannesburg Plan of Implementation, and poverty reduction strategies. However, the success of these initiatives is significantly hampered by a number of barriers. For example, the recent progress report of the MDGs observes that the degradation of ecosystem services and the unsustainable use of natural resources in Africa constitute significant barriers to achieving some of the MDGs by 2015 (United Nations 2007). As such, the adoption of integrated approaches to human development that jointly takes into account ecosystem management and human health concerns is important for meeting the MDGs. Also, in 2005, a draft regional action plan for the implementation of the Strategic Approach to International Chemicals Management (SAICM) was adopted by a number of African countries. The section of the SAICM on human health protection identifies priorities to develop response measures to mitigate environment and health impacts of emergencies involving chemicals, reduce environmental and health risks of pesticides, and ensure occupational health and safety (IISD 2008). Many countries have since been coordinating their efforts towards the implementation of SAICM, and have agreed on regional priorities for health and environment to improve the management of chemicals.

These initiatives indicate that the African region has taken keen interest in preventing the unnecessary deaths caused by environmental factors. A second inter-ministerial meeting is planned for December 2010. This meeting will review progress made on the Libreville Declaration, identify health and environment priorities that need to receive top attention to contribute to achieving the MDGs and also examine the growing concern of climate change on health (UNEP-SAICM 2009).

10.3.2 Health and Environment Policy Frameworks in Europe

As mentioned above, the EU is well advanced with respect to developing and implementing integrated policy frameworks on health and environment, and detailed information about their priorities can be found in various EU documents. As such, we will restrict our discussion to only a few key strategies. Just like many regions in the world, Europeans are concerned about the impact of the environment on their health. These concerns and strategies for addressing them are stipulated in a number of documents, including the *EU Environment and Health Strategy*. The European Commission adopted the Strategy in 2003 with the primary goal to address the links between poor health and environmental problems, and to reduce diseases caused by environmental factors. The Strategy places emphasis on the need to understand and identify health problems related to environmental degradation, so as to be better positioned to prevent new health threats, particularly those linked to pollution. Research is an important component on the EU Strategy.

The *EU Environment and Health Strategy* (2003) is being implemented in successive cycles. The first cycle is being implemented between 2004 and 2010, and focuses on building a good information base on European environment and health issues. It also aims to develop a coordinated approach to Human Biomonitoring by EU member states. An *EU Environment and Health Action Plan 2004–2010* has been developed to implement the strategy. The Action Plan focuses on a number of areas, including understanding the links between diseases and environmental risk factors, and integrating environment and health monitoring and response to gather information and simplify communication between authorities at different levels (Directorate-General for health and Consumers, Health and Environment Factsheet 2010). The Action Plan proposes an integrated approach that involves a closer collaboration among environment, health and research sectors. The goal is to develop a Community System that integrates information on the state of ecosystems, the environment and human health to enable efficient assessment of the environmental impacts on human health.

The second phase of the Strategy commences in 2011, and will focus on strengthening the role of environment and health policy in reducing health inequalities, and responding to the health effects of climate change, among others (ibid).

At the Parma Ministerial conference in 2010, environment and health ministers examined how to better position the region to respond to emerging global challenges within the coming decade. Some of the emerging challenges that EU member states are committed to working on include:

- health and environmental impacts of climate change;
- health risks to children and other vulnerable groups posed by poor environmental, working and living conditions;
- socioeconomic and gender inequalities in the human environment and health, and augmented by the financial crisis;
- non-communicable diseases, especially those that can be reduced through adequate policies in areas such as urban development, transport, food safety and nutrition, and living and working environments;
- endocrine-disrupting and bio-accumulating harmful chemicals and (nano) particles;

In addition to the above, European Union member states acknowledged the importance of making health a central component to socioeconomic development, supported through the development of technologies and green jobs. There was emphasis on integrating health issues in climate change mitigation and adaptation efforts, while also encouraging the integration of environment and health issues into broader strategies and other sector policies (Declaration of the European Commission, Parma 2010). As discussed above, the EU Parliament has endorsed integrating health issues into climate adaptation measures. Other areas identified in the Parma declaration requiring action include: access to safe water and sanitation, equal opportunities for each European child by 2020, improved air quality and an environment free of toxic chemicals, opportunities for physical activity and a healthy diet, and reduction social and gender inequalities (ibid).

In May 2010, European Parliamentarians adopted a resolution that recognizes the important role of environmental factors in cancer prevention. The European Parliament endorsed a report on the European Commission's proposal to establish a European Partnership for Action Against Cancer for the period 2009–2013. This supports EU member states' effort to prevent cancer through environmental policy. Parliamentarians also voted favourably to strengthening health protection in EU climate adaptation policy and acknowledged the possible health benefits that could emerge through adaptation measures.[2] With growing evidence of the role of environmental factors in causing cancer, especially chemical pollution in our everyday environment, there are calls for equal emphasis on the environment as there is on lifestyle factors such as smoking (ibid). The measures adopted by EU are seen by the medical establishment as steps in the right direction and could be emulated by other jurisdictions.

10.4 Developing Integrated Policy Frameworks: Issues and Challenges

There are a number of issues that inhibit the effective integration of environment and health concerns, including limited availability of scientific evidence, limited capacity for policy development, and limited opportunities for advocacy. These challenges manifest themselves differently in different regions of the world. For example, at the global level, there are very few strategic opportunities that avail themselves to advocate the integration of environment, health and development concerns. However, since the Rio Summit, there seem to be increasing international opportunities to draw attention to the growing importance of environmental factors in influencing human health. The recent Climate Change Summit in Copenhagen drew attention to the impact of climate change on a variety of sectors including environment and health. Other global initiatives that have attempted to integrate environment and health concerns with broader human development efforts are the United Nations Millennium Development Goals and the Johannesburg Action Plan. The challenge

---

[2]Health and Environment Alliance website http://www.env-health.org/a/3553?var_mode=calcul. Accessed June 2010

with global initiatives is sustaining the tempo and enthusiasm leading into these events. Prior to these events there is usually a lot of publicity, media coverage and side meetings leading into them. However, once the session is completed, following through with the commitments and translating these into practical programs at the regional, national, and local levels often becomes a challenge. Also, when problems are constructed from a global perspective, they fail to take into account the specific challenges encountered by various population groups, and so end up becoming nothing more than a global event with no human face.

At the regional level, identifying strategic initiatives and opportunities to advance integrated environment and health agendas is particularly important. Regional blocks have to find shared environment and health concerns that are motivating enough to bring countries together to identify joint strategies. With the impending climate change crisis, constructed or not, many regions are exploring ways of integrating health concerns into climate change adaptation policies. This issue is on the agendas of both the EU and the African region as they move forward with their respective objectives on health and environment. In addition, regional ministerial and inter-ministerial conferences provide great opportunities for environment and health ministers to advance, not only regional concerns, but also to identify effective ways of implementing the strategies nationally. These regional gatherings have resulted in the signing and adoption of a number of declarations, including the *Parma Declaration on Environment and Health* by the European Commission and the *Libreville Declaration on Health and Environment in Africa.*

In addition, some regional blocks face the challenge of identifying scientific information linking health and environmental conditions in a timely fashion, to be better prepared to prevent the emergence of new environmental-related diseases. Hence, scientific efforts must aim to provide vital and up-to-date knowledge on health and environment linkages to accurately and timely inform policy.

At the national level, many health and environment ministries still function in isolation. For example, in many African countries, the development of health policies has resided primarily with the ministry of health, seeking little involvement from the environment sector, or sometimes evolving alongside the development of environment policies. Compound ministries such as Science and Technology seem to provide better opportunities for bridging such ministerial silos and seem to allow for the development of broad-based policy that touches on other sectors.

The challenges facing countries, especially in poorer regions, to successfully develop and implement integrated environment and health policies are numerous, and include factors, such as limited capacity to: assess risks and potential health impacts of environmental conditions, collect scientific data, monitor and evaluate the effectiveness of policies and interventions, and translate national health and environment policy into comprehensive action plans that will inform programming at the community level (WHO Regional Office for Africa 2009).

As mentioned above, the differential prioritization of health and environmental issues present challenges for exploring the development of integrated policy approaches. For example, in many African countries, whereas health is usually ranked among the top four in national development strategies, environmental issues do not feature that prominently. Because of this uneven emphasis, the efforts dedicated to improving human health in the public health sector, gets eroded by environmental-related diseases due to the lack of similar efforts in the environment sector (ibid). However, with continuous guidance from UNEP and WHO, and the increasing use of National Environment and Health Action Plans (NEHAPs), many African countries are becoming efficient in integrating health and environment objectives into other national programs. The NEHAPS are government documents that examine environmental health problems from a comprehensive, holistic and inter-sectoral perspective. They are collaboratively developed by a wide range of partners, including technical and professional experts, national, regional and local authorities and nongovernmental organizations (WHO Regional Office for Africa 2009). Also the "healthy settings" approach has been adopted by many African countries as an effective strategy to implementing integrated interventions on health and environment. The ecohealth approach is also gradually being adopted by a number of institutions around the world, including some African countries.

In order to overcome the challenges associated with developing integrated policy frameworks, it is important that governments at all levels acknowledge the importance of the linkages between environment and health, and commit to coordinating their efforts in the two sectors. The health sector must reposition itself and play a lead role in ensuring that health becomes central to all regional, national, and local development policy frameworks. In addition, the health sector must take the lead in coordinating input from all sectors and evaluating the extent to which sector policies such as transportation, energy and housing all take health into account. The private sector is a key player in environmental degradation, yet it plays a limited role in policy development. Efforts must be made to bring the private sector to the table and ensure that private sector activities take into account environment and health implications.

In addition, it is important that regional and national efforts acknowledge research as an essential component of the policy development process. The need for accurate, current, timely and relevant evidence of the links between environment and health is essential. Regions must strive to develop the capacity to collect data on the state, structure and functioning of ecosystems and be able to assess the potential for the emergence of new diseases in an effective manner. Such understanding will allow governments to develop integrated policies that acknowledge these linkages. Policies that are informed by inaccurate or politically tainted science will likely fail to respond to pertinent environmental health threats or even recommend measures that could further augment exposures to environmental health risks. Finally, for developing countries, it is important that they build the necessary scientific and institutional capacity for assessing environment and health linkages, and using that knowledge to develop and implement effective policy tools.

10.5 Influencing Policy Through Community-Based Ecohealth Research

To what extent can research conducted at the community level influence the development of integrated policy at a higher level? Environment and health challenges are probably more prominent at the community level because of the close interaction between people and the surrounding environments. It is also at this level that environment and health concerns are addressed from a disciplinary or sectoral perspective, although there might be an integrated policy framework at the national level. For example, in many developing countries, environmental problems such as land degradation, deforestation, and soil infertility are responded to by extension workers or field personnel from the agricultural, environment and forestry departments, and approached from a departmental and disciplinary perspective. Similarly, community health problems are primarily the responsibility of the health unit, occasionally incorporating environmental health education and primary care dimensions. Yet it is at the community level that the underlying causes of ill health and environmental degradation are so complex and defy any simplistic causal explanations. Environment and health problems at the community level are augmented by socio-economic and poverty concerns. Examining health and environment issues in isolation, and failing to integrate these with poverty reduction strategies, literacy programs, and other broader development objectives is futile.

The ecosystems approach to human health (ecohealth) is particularly useful in providing insight into how these factors interact at the local level to shape environmentally-mediated health outcomes. The chances that the findings of an ecohealth research project could be scaled up to influence broader public policies is enhanced through the use of a transdiciplinary team of researchers or government officials. The involvement of government officials and decision makers from various backgrounds and disciplines allows the problem under investigation to be examined from an integrated perspective, allowing for the development of integrated interventions. For example, previous chapters discussed the findings of an ecohealth project that was conducted in a rural community in Northern Ghana. The project brought together representatives from a number of government departments, including the Ministry of Health, Agriculture, Education, Community Development and Forestry to form a research team and to work in collaboration with local people to identify the major health problems facing the community, examine the underlying causes and develop and implement the necessary interventions (Dakubo 2004). The engagement of various departments in a collaborative investigation at the local level provided an opportunity for the various departments to explore their synergies and come up with integrated approaches to pursuing community development than was previously undertaken.

Sometimes an ecohealth research project at the community level might not influence policy directly and immediately, it can, however, contribute to influencing policy in other important ways, such as altering thinking on health and environment issues, stimulating debates on the use of intersectoral an transdisciplinary approaches, or cause policy makers and their advisors to think critically about the concepts and philosophy of an ecosystem approach to health. Over time then, it can be argued that community-based research has the potential to influence policy indirectly through the circulation of ideas and concepts, and through experimentation of these ideas. Also the involvement of environment and health decision makers as part of the transdisciplinary research team builds the capacity to develop new understanding of environment and health issues and increases the potential to incorporate such ideas into broader policy frameworks. These new ideas also diffuse through society via various means such as conferences, public debates and networking opportunities with decision makers, and eventually find their way up the policy chain.

10.6 Conclusion

This chapter explores a number of integrated policy frameworks at various regional scales that seek to respond to environment and health concerns from a coordinated perspective. While there is increasing awareness of the important role of environmental factors in influencing human health, there seem to be limited capacity and opportunities for developing integrated policies that take these linkages into account. While regional blocks such as the European Union seem well ahead in their ability to respond to environment and health challenges in an integrated manner, regions such as Africa are beginning to build this capacity. Africa is challenged by limited ability to conduct applied research that demonstrates the effects of environmental contaminants on human health. Yet the formulation of effective policy is dependent on sound evidence-based knowledge. There is also limited capacity to efficiently collect, synthesize and interpret technical health and environment data, and monitor and evaluate policies and interventions to assess their effectiveness. Despite these challenges, Africa seems to be making progress with guidance from its leadership - NEPAD, the WHO, UNEP, and Africans in general.

References

Dakubo C (2004) Ecosystem approach to community health planning in Ghana. EcoHealth 1: 50–59

Declaration of the European Commission (2010) Fifth ministerial conference on environment and health: Protecting children's health in a changing environment Parma, Italy, 10–12 March 2010

Directorate-General for Health and Consumers (2010), Health and Environment Factsheet 2010

International Institute for Sustainable Development (IISD) (2008). Health and Environment Bulletin: a report of the first inter-ministerial conference on health and environment in Africa, 26–29 August 2008, Libreville, Gabon

NEPAD (2001) The New Partnership for Africa's Development Policy Framework, October, South Africa

NEPAD (2003). The New Partnership For Africa's Development (NEPAD) Health Strategy: Initial Programme of Action, 2003, South Africa

United Nations. (1993). Earth Summit – Agenda 21. UN. New York, NY, USA

United Nations (2007). The Millennium Development Goals Report, New York, NY

UNEP-SAICM (2009) Strategic Approach to International Chemical Management (SAICM). Update on follow-up to the first inter-ministerial conference on health and environment in Africa.

UNEP/NEPAD (2003). Action plan of the Environment Initiative of the New Partnership for Africa's Development (NEPAD)

WEHAB Working Group (2002). A Framework for Action on Health and Environment. Working Paper for WSSD, Johannesburg, South Africa.

WHO (2002). Resolution AFR/RC53/R3. World Health Organization, Regional Office for Africa, Brazzaville

WHO (2006). Burden of disease statistics. World Health organization. Geneva

WHO Regional Office for Africa (2009). Proceedings from the first interministerial conference on health and Environment in Africa. Libreville, Gabon, 26–29 August 2008

WSSD (2002). A framework for action on health and the environment WEHAB working group, Johannesburg

Part IV
A Critical Approach to Ecohealth Research and Practice

Chapter 11
Applying Critical Theory to Environment and Health Issues

Contents

11.1 Introduction

The next four chapters of this book are designed to map out a critical approach to conducting ecohealth research and explore ways of responding to ecohealth problems through a critical lens. Most traditional literature on environment and health adopts a priori notions of environmental degradation and constructions of poor health without interrogating how these concepts or problems are conceived, who conceives them, why, how, and to what extent they accurately reflect reality? Similarly, the causal explanations offered for environment and health problems are sometimes too simplistic and attributed to factors, such as poverty, poor land use practices, and inappropriate behaviours and lifestyle, with little consideration of how unequal power relations and socio-political and historical factors shape human-environment interactions, and adversely impact health. In addition, how does the mainstream society come to accept scientific environment and health knowledge claims and proposed interventions as legitimate and accurate over other less dominant perspectives?

Such questions underlie critical thought and critical scholars from disciplines such as geography, public health, health geography, and medical anthropology, among others have begun to express concern about the lack of theoretical rigor in the analyses of issues related to environment and health (Baer 1996; Kearns 1996;

Kearns and Joseph 1997; Lupton 1995). Critical scholars are concerned that the failure to adopt a critical perspective to analyzing environment and health problems risk producing solutions and policies that fail to address the underlying causes of ecological and public health problems, or even worse, increase existing social inequities and vulnerabilities (Forsyth 2003).

These concerns have spawned a new breed of critical approaches that aim to develop more politically, socially and ecologically informed explanations to problems. These approaches do not refute the existence of real environment and health problems, rather they draw on critical social theory to develop rigourous analytical frameworks to examine the production of knowledge claims surrounding health and environmental phenomena, the framing of problems and solutions, and the construction of various subject positions (Forsyth 2003; Kearns and Joseph 1997; Lupton 1995). As a new integrated field, the ecosystem approach to human health (ecohealth) stands to benefit from these new theoretical developments. The next four chapters draw on critical perspectives, including poststructuralist theorizing, political ecology, postcolonial, and feminist theories to map out a rigorous critical analytical framework for ecohealth, called *critical ecohealth*.

11.2 Critical Theoretical Perspectives

While the definitions of critical thought vary from one discipline to another, there are some underlying features that are common to this frame of thinking. For the most part, critical thought views knowledge as historically and socially constructed and filtered through perspectives of the dominant society (Foucault 1994; Nicholson 1990). According to this body of literature, especially from poststructuralist perspectives, all knowledges are products of social relations, and change with changing circumstances. What we consider to be "truth" should always be seen or interpreted as the product of power relations, and as such is never neutral or partial, but always acting in the interest of someone (Fox 1994). Poststructuralists believe that there are no objective realities "out there" in the world. They argue that the "truths" that seem to exist have been created by the way we use language. Hence, poststructuralists reject grand narratives of modernist thought, which validate some forms of knowledge as legitimate and morally "correct". For the most part then, critical social theory is concerned about examining the various ways in which dominant knowledge claims are deployed, including how these discourses claim legitimacy, construct the subjects, and propose solutions. They draw on diversity, nuance, and complexity of different experiences to provide myriad ways of understanding phenomena.

Within the field of public health, critical perspectives query the taken-for-granted assumptions underlying public health knowledge and practices, examining who controls these assumptions, and constructs public health problems, as well as attempt to explain how and why alternate views are marginalized (Lupton 1995, 1998). Critical perspectives in public health also seek to illustrate that the root causes of poor health outcomes are embedded in unequal power structures and call for analysis of public health problems to be linked with rigourous political economic and ecological frameworks (Craddock 2000).

Similarly in the environmental sector, the application of critical thought to environmental problems seeks to understand how environmental problems are identified, constructed, explained, and resolved. Critical analytical perspectives of environmental problems seek to illustrate that social and political framings are woven into the construction, explanation and intervention of environmental problems. Just like poor health outcomes, critical environmental perspectives seek to illustrate that the causal factors responsible for environmental degradation are complex, varied, and are rooted more in the power structures of society which shape how people interact with their biophysical environment. Hence a critical approach to ecohealth will draw on critical thinking from both fields to examine problems at the interface of health and environment.

11.3 The Production of Scientific Knowledge Claims

The production of scientific knowledge, including environmental and public health knowledge, has always been associated with providing society with valuable information to make informed decisions or help solve problems. They provide information on how to improve our health and manage our environments, hence persuading society as the legitimate knowledge out there. This legitimacy is partly achieved through scientific backing and the use of specialized language that is readily understood by a few disciplinary experts. In recent years, this dominant scientific knowledge has come under criticism. Scholars drawing on the sociology of knowledge or science studies have begun to question the objectivity and neutrality of scientific truth claims. Science usually sets out to provide an objective explanation of reality, but the experience of this reality is subjective, and is influenced by people's perceptions and interpretations of that reality. From this perspective then, it is helpful to view all knowledge claims as products of social relations, which change with changing contexts. Rather than privilege one particular type of knowledge, critical scholars encourage "plural rationalities", and caution against accepting scientific knowledge claims as "fact" (Fox 1994).

Drawing on Foucault's writing on knowledge and power (1980), critical scholars see the production of dominant knowledge forms to be closely linked with the exercise of power, as such bodies of knowledge have implicit rules on who can speak, from what view point, what can and cannot be said, and in which form (Aviles 2001). Such truth regimes tend to be referred to as the dominant discourse that pervades society and becomes the "common-sense" knowledge. Foucault observes that the centres that produce these knowledge claims are identical to the centres of power (Foucault 1980). Scientific discourses are seen as the outcome of a network of power relations, structures and processes that are able to legitimize certain statements as "truth", while denying others.

Critical scholars make use of discourse analysis to dissect these dominant bodies of knowledge, with the intent to expose any hidden agendas and to contextually explain phenomena. For example discourse analysis of health seeks to illustrate how medical and public health knowledge bodies claim authority, construct and explain health concerns, construct the sick and the vulnerable, and proposes solutions to resolve these health problems. In the field of international development, scholars such as Escobar (1995) and Ferguson (1990) have written about how the discourse of development has been used to produce the Third World as "underdeveloped" compared to the West. By constructing the South as underdeveloped, been development discourse is then able to produce truth claims about health, development and appropriate ecologically practices. There is growing concern that the production of such forms of knowledge has become prolific and hegemonic in nature, and has been deployed in ways that conceal the political, social, and economic interests of the claim makers.

11.3.1 Processes Through which Scientific Knowledge Claim Authority

In order to remain dominant and maintain power, scientific discourses must be seen as legitimate, and such legitimacy and authority are maintained through a number of processes that are not readily apparent to society. Critical scholars draw on discourse analysis to examine and reveal the ways through which dominant knowledge claims gain legitimacy and maintain authority. Discourse analysis examines how texts and language are deployed within socio-cultural practice. It focuses on how identities, social relations, and knowledge are constructed in spoken and written texts. In particular, discourse analysis focuses on the claims-making process, the claims concerning the phenomenon, the claims-makers themselves, and the social impacts and policy outcomes of those claims (Hannigan 1995). Discourse analysis seeks to answer questions such as: What power strategies do these claims engender? What perspectives, issues and questions get silenced, disguised, or eliminated in the production and circulation of dominant knowledge claims.

Scientific discourses gain and maintain legitimacy through the use of specialized languages. Because everyone in society does not readily understand this language, power is automatically conferred to the selected few who are able to speak this specialized language. The formal scientific training and expertise acquired by these selected few further confers legitimacy in them to explain phenomena such as climate change from a scientific perspective. Boundaries are then set between expert and lay, with the expert's knowledge claims having more weight than the layperson, and hence being seen as legitimate and adopted and implemented by institutions. Scientific discourses also achieve validation through the construction of problems and those experiencing the problems. Problems are constructed as deviating from the "norm" which is established by scientific discourse. Constructed as deviating from the norm, scientists are then justified to intervene and bring it back to conform with the norm (Lee and Garvin 2003). Abnormalities are denoted by constructing practices and behaviours as "inappropriate", as in inappropriate land use practices or health behaviours. Hence without the constructions of abnormalities, interventions cannot be justified. Also, the interventions that emanate from dominant discourses and the institutions that adopt them often appeal to the "social or common" good. By subscribing to socially accepted ideologies, institutions and professionals render "natural" their subordination of other forms of knowledge, while denying or concealing any political or economic interests (Escobar 1995). For example, with respect to natural resource management, the state and other resource management institutions propose interventions that constrain people's use and access to resources, while legitimizing their practices as "scientifically proven" and hence "ecologically good" versus the "ecologically bad" practices of farmers and peasants (Schmink and Wood 1987). Similarly, within the health sector, public health discourse legitimizes the dominance of medical knowledge over lay knowledge, and portrays the health professional as the expert. The recommended practices of public health professionals tend to be associated with the public good, while attention is being drawn to the inappropriate health practices that ordinary citizens engage in.

11.4   Constructing Subject Positions Through the Use of Binary Logic

As discussed above dominant discourses claim authority by constructing certain phenomena as appropriate versus inappropriate, developed versus underdeveloped, and lay versus experts. Such dualisms draw on binary logic to construct one practice as normal and the other as abnormal. Similar notions have been used to construct subject positions. People are seen as being constituted in and through discourses and social practices. For example, in the context of health, how do people come to perceive of themselves as "healthy" or "sick"? How do health professionals come to conceive of people as "sick" or "healthy"? Is there a universal experience of "healthiness" or "sickness"? Who defines health, and for whom?

Critical scholars have called for the need to interrogate science for its cultural constructions of human subjects (Jasanoff and Wynne 1998). In particular, feminist scholarship has criticised the use of binary logic in scientific constructions. They argue that by dividing events into two opposing categories, we assume that there is some intrinsic opposition between the two; one category contains similar elements, compared to the other category. For example, in the context of gender, we divide society into men and women, and assume that women have more in common with each other than they have with men. This premise creates a false appearance of unity by reducing the difference and heterogeneity of experience (*of men and women*) into supposedly natural or essentialist oppositions (Flax 1990: 36). By dividing events into two opposing categories, there is the tendency to continue to add a series of distinctions to the initial two categories, such as men are healthy, women are unhealthy;

men are rational, women are irrational; women are more vulnerable, men are stable, and so on. Feminist scholarship argues that it is through the use of binary logic that we privilege one category over another; we privilege the male body as healthy, and the female as unhealthy (Annandale and Clark 1996). Critical scholars therefore argue that scientific constructions of various subject positions such as, sick and healthy, men and women, lay and expert, are discursive categories created through the use of binary logic (Fox 1994).

Critical approaches aim to unravel such binary logic and to reveal them as particular ways of knowing the world (Fox 1994). They argue that there are no fixed oppositions, as reality is constantly shifting, and socially constructed out of competing discourses. Hence, in a heterogeneous postmodern social world, there is no single "truth", but a series of multiple "truths". Instead of conceiving of people as expert or lay, we should try to illustrate how the dominant privileged position (expert) is created in contrast with the weaker position (lay). This means that we can only conceive of expert by conceiving of who is not an expert, and "developed" by conceiving who is not "developed". This therefore reveals that what we think of as opposites are not necessarily opposites, but are interdependent, constructed out of each other, for their own authenticity (Grosz 1990). The same logic is used with nature/society debates, as nature and society co-constitute each other and hence nature cannot be conceived in isolation of society. This is a central element of ecohealth whereby human beings are perceived as integral to the ecosystem.

Following this then, feminist poststructuralists have called for different ways of investigating gender-related issues. They ask that we dislodge the opposition between men and women, avoid oversimplification and instead, try to contextualize the experience of both men and women's lives as they are embedded in place, class, socio-economic status, and ethnicity (Annandale and Clark 1996). The focus should be on the complexities of people and their specific situations and experiences at one particular point in time, and not on developing concepts based on abnormalities, similarities and differences. In a heterogeneous world, where individuals possess multiple, fractured, overlapping, and changing identities (Butler 1992), it is difficult to try to reduce such varied experiences into one category, such as "woman."

It is important for scientific constructions to perceive the subject as fragmented and constituted in difference, and should not be constituted through essentializing, universalizing and generalizing concepts. The concept of uniformity risks smoothing over differences, conflicts of interests, and contradictions inherent in people's everyday lives. They also risk taking events out of historical context and dynamism, and creating illusions of homogeneity, coherence, and timelessness (Abu-Lughod 1993: 9). Some scholars suggest that generalization of people's experiences by professionals is an expression of power (ibid). It is a language of powerful actors who seem to stand apart from, and outside of what they are describing. The distinction between professional discourses of generalization, and the languages of everyday life creates a boundary between the knowledge constructor on one hand, and the constructed on the other (ibid).

11.5 Scientific Interventions and Forms of Resistance

Scientific knowledge claims serve as the basis for the development of interventions, policy and decision-making. However, when interventions are developed based on uncritical science, they fail to address the underlying causes of the problem under investigation. In addition, there is always an assumption that those affected by the problem will uncritically accept the proposed interventions. But as Scott (1990) points out, intended beneficiaries are able to challenge dominant knowledge claims through a culture of everyday resistance and avoidance behavior, thereby undermining the dominance of the discourse. For example, within the spheres of natural resource management, everyday resistance takes the form of illegal harvesting and poaching of wildlife, encroaching into protected areas, or through conducts of arson (Marchak 1995; Peluso 1992; Scott 1990). In the health sector, dominant knowledge systems may be resisted through avoidance behaviours and non-compliance with health education programs and so-called 'appropriate' health behaviours and practices. These resistance activities challenge the "social" and "common" good notions of interventions flowing from dominant discourses, thereby highlighting the agency and adaptive powers of ordinary citizens.

In addition, Foucault suggests that the production of dominant knowledge claims does not manifest itself through simple bipolar relations of who has power and who does not have power. Instead, relations of power manifest themselves differently in different cultures, and are organized through structures such as class, age, religion, marital status, and education. Foucault posits that alternative perceptions and forms of knowledge can resist dominant knowledge systems, consciously or subconsciously. These alternative forms of knowledge are produced at individual and group levels and their collective power is enhanced through sharing and convincing others to accept these forms of knowledge as valid ways of knowing (Weedon 1987). Foucault therefore encourages the acceptance of multiple centres of truth and the incorporation of subjugated local knowledges of indigenous people into the decision-making process, thereby enhancing the acceptability of interventions.

11.6 Conclusion

This chapter outlines some of the key elements of critical social theory. It draws attention to the production of scientific knowledge claims and the various ways through which such dominant discourses gain legitimacy and maintain authority. It also discusses how subject positions are constituted through the use of binary logic, which reduces the flux and heterogeneity in people's lived experiences. Finally, the chapter draws attention to the fact that dominant discourses and the interventions they propose can be resisted through ways that are not readily apparent to the scientific professional. In the next two chapters, we will examine how these critical perspectives are deployed in environment and health discourses, before exploring how they might help inform ecohealth as a discipline in the final chapter.

References

Abu-Lughod L (1993) Introduction. In: Abu-Lughod L (ed) Writing women's world. Bedouin stories. University of California Press, Berkeley, CA, pp 1–42

Annandale E, Clark J (1996) What is gender? Feminist theory and the sociology of reproduction. Sociol Health Illn 18(1):17–44

Aviles LA (2001) Epidemiology as discourse: the politics of development institutions in the Epidemiological Profile of El Salvador. J Epidemiol Commun Health 55:164–171

Baer HA (1996) Bringing political ecology into critical medical anthropology: a challenge to biocultural approaches. Med Anthropol 17(2):129–141

Butler J (1992) Contingent foundations: feminism and the question of "postmodernism". In: Butler J and Scott JW (eds) Feminists theorize the political. Routledge, New York, NY, pp 3–21

Craddock S (2000) Disease, social identity, and risk: rethinking thegeography of AIDS. Trans Inst Br Geogr NS 25:153–168

Escobar A (1995) Imagining a post-development era. In: Crush J (ed) Power of development. Routledge, London, pp 211–227

Ferguson J (1990) "The anti-politics machine: "development," depoliticization, and bureaucratic power in Lesotho". Cambridge University Press, Cambridge

Flax J (1990) Thinking fragments: psychoanalysis, feminism, and postmodernism in the contemporary west. University of California Press, Oxford

Forsyth T (2003) Critical political ecology: the politics of environmental science. Routledge, London

Foucault M (1980) Power/knowledge. Pantheon, New York, NY

Foucault M (1994) The birth of the clinic: an archeology of medical perception. Vintage.

Fox NJ (1994) Postmodernism, sociology and health. University of Toronto Press, Toronto, ON

Grosz E (1990) Contemporary theories of power and subjectivity. In: Gunew S (ed) Feminist knowledge: critique and construct. Routledge, London

Hannigan JA (1995) Environmental sociology. Routledge, London

Jasanoff S, Wynne B (1998) Science and decision-making. In: Rayner S, Malone EL (eds) Human choice and climate change, vol 1: the societal framework. Batelle Press, Columbus, OH, pp 1–87

Kearns R (1996) Medical geography: making space for difference. Prog Hum Geogr 19:251–259

Kearns R, Joseph AE (1997) Restructuring health and rural communities in New Zealand. Prog Hum Geogr 21:18–32

Lee RG, Garvin T (2003) Moving from information transfer to information exchange in health and health care. Soc Sci Med 56:449–464

Lupton D (1995) The imperative of health. Public health and the regulated body. Sage, London

Lupton D (1998) The emotional self: a socio-cultural exploration. Sage, London

Marchak P (1995) Logging the globe. McGill-Queen's University Press, Quebec

Nicholson L (ed) (1990) Feminism/postmodernism. Routledge, New York, NY

Peluso NL (1992) Rich forests, poor people: resource control and resistance in Java. University of California Press, Berkeley, CA

Schmink M, Wood CH(1987) The "political ecology" of Amazonia. In: Little PD, Horowitz MM (eds) Lands at risk in the third world: local-level perspectives. Westview Press, Boulder, CO, pp 38–57

Scott JC (1990) Domination and the arts of resistance. Yale University Press, New Haven, CT

Weedon C (1987) Feminist practice and poststructuralist theory. B. Blackwell, Oxford, UK

Chapter 12
Examining Environmental Problems from a Critical Perspective

Contents

12.1 Introduction

Since the Earth Summit in Rio, there have been growing concerns about the deteriorating state of the world's environment and role human activities play in this transformation. Problems such as climate change, deforestation, biodiversity loss, and desertification have become prevalent. The Millennium Ecosystem Assessment (MA 2005) suggests that over the past half century, human activities have changed the natural ecosystem more rapidly and extensively than in any time in history. As a result, close to 60% of the world's ecosystem services are being degraded or used unsustainably (ibid). The consequences of a degraded environment are enormous as they affect livelihoods options, predispose people to new infectious diseases, and cause more unstable weather events.

While these environmental problems are real and will likely impact society, especially vulnerable groups, the circumstances surrounding the identification, explanation, and resolution of these problems are not clear and are sometimes attributed to wrong causal factors. For example, environmental problems such as land degradation and deforestation, especially in the global south, tend to be attributed to rapid population growth, poverty, and inappropriate land use practices. In the North, consumerism, maldevelopment and industrial pollution seem to be the main culprits. These explanations are based on one particular environmental discourse, the dominant discourse that sustains society. While there might be other

C.Y. Dakubo, *Ecosystems and Human Health*, DOI 10.1007/978-1-4419-0206-1_12,
© Springer Science+Business Media, LLC 2011

explanations or environmental discourses, they tend to be silenced or fail to receive the stamp of approval as legitimate knowledge due to lack of scientific backing.

Most often than not, it is the dominant scientific explanations that are adopted as the standard knowledge, with Indigenous knowledges and others complementing and filling the gaps. As discussed in the previous chapter, the production of scientific knowledge claims have come under scrutiny for failing to contextualize problems within socio-political, historical, and cultural contexts. The failure to contextualize environmental phenomena may produce "facts" or result in explanations that do not reflect biophysical and social realities. In the event that these "facts" are used to develop interventions or inform policy, there is an increased likelihood that these interventions could fail to respond to the problem at hand or result in policies that unfairly penalize land or resource users (Forsyth 2003). This points to the importance of ensuring that environmental problems are diagnosed accurately, prior to prescribing any treatment. Just like medicine, a wrong diagnosis will consequently lead to a wrong treatment or prescription, which if implemented, may worsen the patient's condition or lead to death.

By calling for the contextualization of environmental problems, critical scholars, including poststructuralist political ecologists seek to illustrate that the construction of some environmental problems are not neutral, but instead are shaped by the political and social interests of the claim makers. In particular, critical scholars argue that the metaphorical description of many environmental problems such as in "global crises", and "extensive degradation" should not be accepted uncritically as fact. The concern is that by framing problems from a global perspective, there is a tendency to prescribe global solutions which are unable to respond to specific environmental problems of various population groups. Some of these global interventions propose universal standards of ecological conservation that obscure heterogeneity and conflict, and present environmental problems as easily molded for uniform solutions (Goldman and Schurman 2000). Critical scholars are particularly cautious of global conservation programs and knowledge systems, including those environmental organizations that purport to work for the social good (Goldman and Schurman 2000; Peluso 1993; Schroeder 1995). In addition, critical scholars seek to understand the power dynamics surrounding the production of Southern environmental knowledge claims by Northern scholars and institutions.

Since the 1990s, a number of alternative discourses on environmental problems have emerged, with some seeking to understand how environmental problems are conceived, explained and resolved. The emergence of these alternate discourses was driven in part by some seminal environmental research that was conducted in developing countries on issues such as deforestation, desertification and soil erosion. The findings from these studies suggested that explanations offered for many environmental problems were not historicized, and were based on outdated, and partial and inaccurate accounts of environmental degradation (Batterbury et al., 1997). Such works also demonstrated how environmental problems in developing countries were not the result of short-term impacts of rising population growth, but instead the result of complexities resulting from long-term human-environment interactions (Morse and Stocking 1995). For example, two of these studies, *The political economy of soil erosion in developing countries* by Blaikies (1985) and *Uncertainty on a Himalayan scale* (Thompson et al., 1986) refuted explanations about the degrading impacts of deforestation and soil erosion in the Himalayas. In addition, Taylor and Buttel (1992) used the Limits to Growth study and concerns about global climate change to illustrate that "politics are woven into environmental science at its 'upstream' end" (p 406). Also, in their study on *How do we know we have global environmental problems?* Taylor and Buttel argue that the construction of environmental problems from a global perspective

> involves a universalizing discourse that steers us away from the difficult politics of enduring structural inequalities and differentiated interests and toward technomanagerialist remedies, preferred (and constituted) by elite, Northern-based scientists and bureaucrats. (cited in Goldman and Schurman 2000: 575).

Following the climate change conference in Copenhagen in 2010, many critics have interrogated the knowledge claims and magnitude of the impending crisis espoused by climate change scientists. While the debate on the validity of scientific data on climate change and its impending impact is on-going, some scholars caution that it will be at society's peril to dismiss the existence of real environmental problems, including those of climate change as crisis-driven environmental "orthodoxies" fiction or constructions of the North (Batterbury et al., 1997). Instead the objective should be to develop environmental explanations that are socially and politically aware, yet being aware of the existence of real biophysical and climatic issues.

Some critical scholars are also concerned about illustrating how explanations for environmental problems, especially those related to ecosystem degradation, should be examined from the perspective of unequal power relations in society and how this controls access to, and use of ecological resources. Instead of viewing the land user or ordinary citizen as the perpetrator of environmental degradation, explanations should broaden to include the state extractive, industries their capitalist expansion activities, and environmental policies that tend to constrain livelihood options thereby forcing land users to engage in deteriorating practices.

12.2 Examining Various Environmental Discourses

As mentioned above, there are a number of environmental discourses currently in circulation, with some being more dominant than others. The various environmental discourses vary based on what or who they perceive as the primary drivers of environmental degradation, who they perceive to be the victims of such environmental degradation, and what they consider to be appropriate solutions. For example, Adger et al. (2001) examined three circulating environmental discourses – *global environmental management (GEM)*, *populist*, and *denial* discourses – that were used to construct, explain, and prescribe solutions for four environmental problems, including climate change, desertification, deforestation and biodiversity. The *global environmental management* (GEM) discourse identifies the driving factors of environmental degradation to comprise of poor land use practices of local actors and rapid population growth. It also constructs environmental problems as global crisis and draws on scientific evidences such as the Intergovernmental Panel on Climate Change to substantiate the claims. Most of the suggested solutions are global in nature, and calls for the development of coordinated multi-lateral policy frameworks and international actions. In addition, solutions are mostly top-down, tenchnocentrist and involved North-South technology and resources transfers.

The second discourse, the *populist* discourse, concurs with the construction of the environmental problems under discussion as "global" in scope, it however differs with respect to the factors driving environmental degradation. For example, while the GEM discourses see local people and resource users as perpetrators of environmental degradation, the populist discourse identifies globalization, uneven terms of trade and global capitalism as the underlying culprits of environmental degradation. The populist discourse sees local actors as victims reckless environmental activities by external agents. Populist interventions focus on self-determination, social justice, and empowering people to take action to control environmental problems. They oppose external interventions, instead advocating for the use of traditional knowledge as the basis for sustainable practices.

The third discourse discussed by Adger et al. (2001) is the *denial* discourse. Denial discourses interrogate the existence of certain environmental problems, and questions the framing of the magnitude, and causal explanations. A typical area where denial discourses are prominent is climate change, with many critics challenging the science and data of climate change. Denial discourses seek to challenge the objectivity and political neutrality of the construction of environmental problems. They also seek to overturn the explanations offered for the emergence of environmental problems, instead seeking to explain them based on political, historical, economic, and institutional contexts. For example, Fairhead and Leach (1998) drew on denial discourses to overturn explanations offered for deforestation in a West African country. The presence of "relic forests" in the community was erroneously taken as evidence of extensive deforestation. However, drawing on historical accounts of forest cover change, Fairhead and Leach (1998) were able to demonstrate that the "forest relics" were not signs of deforestation, but instead were islands of human induced forest savanna. The study overturned the findings that identified land users as destroyers of forests to preservers of forests, and put to doubt other information and knowledge claims related to global environmental change and forest depletion.

Denial and populist discourses have spawned interest in broadening investigation of environmental problems beyond scientific procedures to incorporating elements of critical social theory and how this provides a better understanding of people-environment relationships. This critical lens has resulted in "new" ecological thinking that draws on Foucault's writing on knowledge and power, and the sociology of science to situate the construction of environmental problems within various social and political regimes (Leach and Mearns 1996).

Political ecology is one of those frameworks that adopts a critical approach to environmental problems, by examining how capitalist activities and the uneven distribution of power influence people-environment relationships (Blaikie and Brookfield 1987; Bryant and Bailey 1997). Political ecology also draws on poststructuralist perspectives to examine the discursive formations of environmental knowledge claims and the practices and policies they engender (Peet and Watts 1996). These two streams of political ecology are sometimes referred to in the literature as "old" and "new" political ecology, with the latter sometimes referred to as "critical" or "poststructuralist" political ecology (Escobar 1996; Forsyth 2003). While political ecology proves to be a useful analytical framework, its application in the field of ecohealth has been limited.

12.3 Key Features of Political Ecology

According to Blaikie and Brookfield (1987: 17), "the phrase 'political ecology' combines the concerns of ecology and a broadly defined political economy. Together this encompasses the constantly shifting dialectic between society and land-based resources, and also within classes and groups within society itself." This definition focuses on the negotiations that occur between land users and other actors relating to access and use of land-based resources. Other scholars, refer to "political ecology" as the politics of environmental problems, with little reference to "ecology". For example, Bryant (1992: 13) describes political ecology as an inquiry into "the political forces, conditions and ramifications of environmental change."

Political ecologists see the environment as a "politicized" space in which environmental problems are simultaneously political-economic problems and so cannot be examined outside their political and economic contexts (Bryant and Bailey 1997). A politicized environment is constituted through unequal power relations between various actors, and political ecologists usually are interested in examining how this unequal distribution of power is used to: (a) control access to, and use of environmental resources; (b) control the distribution of the costs and benefits associated with environmental activities; and (c) control the selective identification, prioritization and representation of environment problems (Bryant and Bailey 1997; Bryant 1998).

Power is seen as a key concept in specifying the various dimensions of a politicized environment. According to political ecologists (Bryant and Bailey 1997: 39), power is the ability of an actor to control their own interaction with the environment and that of others. Political ecologists adopt an inclusive understanding of power to encompass the material and non-material dimensions of power as well as the apparent fluidity of power. To appreciate the role that power plays in shaping human-environment interaction, it is important to examine the various ways and forms in which can actor a powerful actor control the environment of weaker actors, and how such relations are inscribed on the environment and resisted by weaker actors.

According to Bryant and Bailey (1997), there a number of ways through which power can be exerted over the environment of weaker others. One such means is to control access of other actors to a diversity of environmental resources, such as land, water, forests, non-timber forest products. Usually such control is done with the intent to monopolize the economic and ecological benefits associated with the resource in question, to the detriment of weaker actors. Control may take a variety of forms including the use of policies and legislation. For example, in many parts of Africa, both colonial and post-independence authorities controlled and continue to control access to vital natural resources through policies of total and partial exclusion, and through the establishment of game and forest reserves (Bryant 1997; Peluso 1992). Hence, to the extent that one actor is able to control who exploits what resources, where, how, under what conditions and for what purposes, then they have succeeded in exercising their power over the environments of other actors.

Given the vitality of natural resources to the livelihood of many communities, especially in developing countries and Aboriginal communities, control over access to resources further marginalizes land users, limits their livelihood options, and renders them vulnerable to the effects of environmental change. Also, by virtue of their marginal political and economic status, weaker actors tend to inhabit areas prone to extreme weather events due to climate change or pushed onto marginal lands. In order to continue to sustain their livelihood, land users have no choice but to intensify their interaction with the biophysical environment, making use of intensive production systems and in the process degrade ecological resources or render the land less productive, which further constrains their source of livelihood and health. As Blaikie and Brookfield (1987) observe this cycle of environmental degradation then becomes both a result of and a cause of social marginalization. Also, not only are weaker actors trapped in this vicious circle with few options, but they also are accused of degrading the environment through inappropriate land use practices. Ironically, the social and economic marginalization of weaker actors further strengthens the political control of powerful actors, as they continue to maximize the economic benefits of their activities, while deflecting the adverse impacts to other people and environments (Bryant and Bailey 1997).

The control over access to environmental resources is closely linked to a disproportionate distribution of the costs and benefits associated with environmental disaster problems. What is usually perceived as an environmental resources and for weaker actors, turns out to be an economic opportunity for powerful actors, as seen in the extraction of mineral resources in many Indigenous communities in North and Latin America. In most cases, it is the weaker actors who bear the disproportionate burden of local environmental problems that are related to the economic activities of powerful actors (states, businesses, and multilateral institutions). In addition to being disproportionately burdened, they also have little resources to cope with or escape the costs associated with the environmental problems. In contrast, powerful actors consolidate their position in society partly from the economic activities related to the environmental problem. They also have the ability and necessary resources to escape, displace or mitigate the associated environmental effects (Bryant and Bailey 1997). This undifferentiated impact further reinforces marginalization and leaves communities with fewer options.

Power is also exercised through control over the societal prioritization of environmental projects and problems. Powerful actors such as the state, and global institutions are able to identify what environmental issues should be prioritized where, at what time, and how much resources should be committed. Critical scholars argue that, this selective identification and prioritization of environmental problems is itself a political process and may or may not be grounded in scientific "fact" or existing reality, but instead linked to the political and social interests of powerful actors (Bryant 1998:88). For example, during the colonial era, many African indigenous land use practices were blamed for causing soil erosion, prompting the implementation of corrective soil "conservation" measures that either regulated farmers' land-use practices or displaced farmers from so-called "threatened" areas (Neumann 1996). Although these colonial discourses on soil erosion were not informed by any sound explanations, they still persist today and have remained popular with mainstream scholars and policy makers (Bryant 1998). The institution of environmental interventions derived from uncritical explanations have led some to question the efficacy and neutrality of these interventions. According to Guthman (1997: 45), the production of environmental interventions is closely linked to the production of environmental knowledge, both of which are intrinsically bound up with power relations, and hence not neutral. Hence, interventions that emerge from such partial reconstruction of environmental phenomena could end up reinforcing social and economic inequities to the extent that these knowledge claims are used to develop public policy (Forsyth 2003).

This latter point illustrates that power is not only exercised through the control of material resources, but also through the regulation of ideas and the control of environmental discourse (Schmink and Wood 1987). For example, power is exercised through the ability of an actor to define an environmental problem, prescribe the appropriate solutions, delineate what is accepted as "appropriate" environmental discourse or practice, or construct and interpret subjective positions and their experiences. This new direction, responds, in part, to Peet and Watts' (1996) concern that environmental problems should not only be examined from a political-economic perspective, but should be combined with the discursive and ideological realms to reveal how constructions of nature and politics interact to shape material reality. This discursive turn draws on poststructuralist perspectives and is sometimes referred to as "poststructuralist" or "critical" political ecology.

The goal of critical political ecology then is to examine the ways in which knowledge and power interrelate to mediate ecological outcomes and the explanations offered for environmental degradation (Escobar 1996; Peet and Watts 1996). Critical political ecologists argue that the construction of environmental knowledge claims is not value-free or politically neutral. They caution against the adoption of pre-existing notions of environmental problems, concepts and causal explanations as accurate, without evaluating them within the social and political contexts surrounding their framing. They challenge "global" constructions of many environmental problems and their associated global solutions, arguing that such constructions elide heterogeneity and conflict (Goldman and Schurman 2000). Critical political ecology draws on discourse analysis to explore the mechanisms through which environmental knowledge claims are produced, gain legitimacy, and propose socially good interventions. Through this analysis critical political ecology seeks to find answers to questions such as: How do particular environmental knowledges become so pervasive and constitute the "common-sense knowledge" that sustain society? What is the relationship between those who create this knowledge and the rest of society? Why do certain knowledges get privileged and others silenced? How are the facts contested (Peet and Watts 1996)?

Environmental discourses proposed by powerful actors, including the state and international organizations become dominant forms of knowledge that sustain society. These institutions and the dominant discourses they subscribe to, produce interventions that are "ecologically good" compared to the "ecologically bad" practices of weaker actors. To achieve legitimacy, these interventions appeal to the "common good" (Schmink and Wood 1987) and draw on scientific evidence to support the production of environmental interventions as seen in the GEM discourse above. By subscribing to scientifically proven techniques, powerful actors are able to render neutral any political and economic interests or their subordination and marginalization of weaker actors (Escobar 1996).

Critical political ecology, therefore argues that ideas, environmental knowledge claims, and interventions are never innocent, but act to either reinforce or challenge existing social and economic arrangements. It challenges notions of human-induced environmental crisis and their associated technical solutions, and instead argues for the centrality of unequal power relations in shaping human-environment interaction, both materially and discursively (Blaikie 1995; Bryant and Bailey 1997).

In addition to the above issues, there are a number of conceptual elements that inform political ecology. For example, political ecology examines how interaction of phenomena at different scales influences local environmental degradation. It illustrates how environmental problems at the micro-level (household level) are shaped by broader macro-level structures at the community, state, national and international levels, including national and international policies (Blaikie 1995; Bryant 1997). Like the ecosystem approach, it makes use of the concept of the nested hierarchy of analysis, which illustrates how individual health is shaped by the interaction of factors from various levels, including the family, community, nation-state, and the biosphere. The recognition that ecological and social systems interact unevenly across time and space, lead to the understanding that human-environment relations are not static, but always changing with various circumstances. This dynamic nature also illustrates that environmental experiences are not uniform nor are they always generalizable (Harper 2004).

As discussed above, political ecology calls for the contextualization of human-environment relationships, which implies a focus on the actor and the social relations in which they are embedded either at the micro- or macro-level. For example, how do variables such as gender, age, socioeconomic status, and ethnicity mediate access to, control over, and use of resources at varying scales and time? Some scholars argue that society-environment relations are shaped by gender. For example, power relations between men and women are manifested through the control over access to various environmental resources, and through inequitable distribution of environmental rights, responsibilities, benefits and costs (Carney 1996). Gender power relations over environmental resources are also reflected in the differential income-earning opportunities between men and women, and how women's household activities fail to be recognized as "work" (Agarwal 1997; Joekes et al., 1995; Rocheleau et al., 1996). Hence, gender relations, including power struggles in the household are fundamental to understanding access to, use and control of resource and the processes leading to environmental degradation (Agarwal 1992, 1994; Joekes et al., 1995; Leach 1991, 1994). They also illustrate how gender-patterned interaction with the biophysical environment produce gendered knowledges of agroecological systems (Mackenzie 1995; Rocheleau 1995). These gendered environmental knowledges further cast doubt over generalized and universal definitions and experiences of environmental degradation. Using gender as an example, it is important that environmental interventions pay attention to the axes of difference in society and how these mediate resource use. The failure to include these in our analysis of people-environment relations could result in measures that risk increasing environmental inequities, including gender marginalization to resource use.

Finally, political ecology emphasizes the importance of situating environmental transformations within an historical context, so as to understand how past events shape present conditions. For example, in the study by Fairhead and Leach (1995), the incorporation of "events history" in their analysis led to the reversal of explanations regarding deforestation, thus challenging the commonly perceived causes of deforestation. Political ecologists are therefore concerned that by drawing solely on scientific data and quantitative procedures, the construction of environmental problems may fail to adequately reflect the political struggles and historical antecedents that interact at various scales and times to influence environmental degradation, or fail to reflect traditional knowledges systems of environmental degradation.

The application of political ecology to ecohealth concerns provides a comprehensive, integrated framework from which to better appreciate how unequal power relations shape ecosystem change and produces ill health. The application of political ecology perspectives to human health concerns is referred to as the political ecology of health, and it is examined next.

12.4 Political Ecology of Health

The application of the principles of political ecology discussed above to human health concerns allows for an understanding of how political and economic processes shape human-environment interactions and how such interactions contribute to disparate environmental health risks, exposures and health outcomes (Kalipeni and Oppong 1998; Mayer 1996, 2000). In other words, the examination of health problems from a political ecology perspective requires that we link the contexts in which human-environment interactions occur, the differential benefits and costs they engender, and how these are distributed, with rigorous political economic and ecological frameworks. A political ecology approach to health attempts to connect large-scale, as well as micro-scale political, economic and social processes to local health experiences.

Mayer (1992) suggests that the underlying cause of poor health outcomes is rooted as much in social structures and relations as it is in microbes. Hence, those environmental conditions that produce ill health are seen to result from a complex interplay of socio-ecological and political factors that are derived from the uneven power relations that characterize people-environment interactions. Access to, and use of, health enhancing environmental resources are further constrained by such unequal power relations, thus providing a means for understanding the spatial and social disparities in health outcomes that characterize various population groups and communities (Harper 2004).

From a political ecology perspective, then, the concept of causality and the determinants of ill health must be broadened to encompass social, political and economic factors, and must examine how the interaction of these factors leads to ecological change and subsequently to malnutrition and epidemics. As Turshen argues "most analyses separate ecological change from malnutrition, political struggle from epidemics, and social aspects of disease from economic transformation" (Turshen 1984: xi). To the extent that health problems are simultaneously political economic and ecological problems, interventions must be broadened beyond the health and environment sectors to include power dynamics and social organization.

With a few exceptions (e.g. Mayer 1996, 2000), the application of political ecology to health and healthcare issues is limited, although some tenets of political ecology have been applied elsewhere. For example, Hughes and Hunter (1970) examined the connection between large-scale development projects (specifically dam construction) and increased incidence of diseases such as malaria and schistosomiasis in the Upper East Region of Ghana. This study emphasized the importance of understanding disease within the dual frameworks of modernization and sociopolitical developments. Turshen's (1984) work examined the impact of global, social, and economic forces and the impact of historical factors on endemic diseases in Tanzania, such as, trypanosomiasis. Turshen's work contributes to a better understanding of disease occurrence in the colonial context. Other studies include Kalipeni and Oppong's (1998) study on the application of political ecology to examining the circumstances underlying the refugee crisis in Africa; and Hunter (2003), used political ecology to explore the relationship between agricultural development and the incidence of urinary schistosomiasis in the Upper East Region of Ghana. Within the context of ecohealth, the use of political ecology of health is very limited. Some initial applications were by Dakubo (2004, 2006), who applied critical political ecology to an ecohealth research project in Northern Ghana.

12.5 Conclusion

This chapter draws on political ecology framework to illustrate the application of critical perspectives to environmental concerns. Political ecology allows us to examine how knowledge and power interrelate to mediate people-environment relations, including ecological outcomes and explanations. When applied to health issues, a political ecology of health allows for a better understanding of the interplay of socio-ecological and political factors in shaping health determinants and the socially and spatial patterning of various health outcomes. The chapter also examines the features of common environment discourses, including the global environmental management, populist and denial discourses. While global managerial discourses construct environmental problems from a global perspective and attribute the causes to land users, the populist discourse sees globalization and capitalist forces as the main drivers of environmental degradation, with local actors being the victims. Denial discourses, on the other hand, question the processes leading to the constructions of environmental problems as global crisis, and seek to contextualize the occurrence of environmental problems. With increasing concern about the role of environmental degradation on human health, it is important that we make use of rigorous analytical frameworks that have the ability to explore and expose the complexity of issues informing human-environment interactions.

References

Adger WN, Benjaminsen TA, Brown K, Svarstad H (2001) Advancing a political ecology of global environmental discourses. Dev Change 32:681–715

Agarwal B (1992) The gender and environment debate: lessons from India. Fem Stud 18:119–158

Agarwal B (1994) A field of one's own: gender and land rights in South Asia. Cambridge University Press, New York, NY

Agarwal B (1997) Environmental action, gender equity and women's participation. Dev Change 28:1–43

Batterbury S, Forsyth T, Thompson K (1997) Environmental transformations in developing countries: hybrid research and democratic policy. Geogr J 163(2):126–132

Blaikie PM (1985) The political economy of soil erosion in developing countries. Longman Scientific, London

Blaikie P (1995) Understanding environmental issues. In: Morse, S and Stocking M (eds) People and environment University College London Press, London, 1–30

Blaikie P, Brookfield H (1987) Land degradation and Society. Metheun, London

Bryant RL (1992) Political ecology: an emerging research agenda in third world studies. Polit Geogr 11:11–36

Bryant RL (1997) Third World political ecology: An introduction. In: Bryant, RL, and S Bailey (ed) Third world political ecology, Routledge, London ; New York, NY

Bryant RL (1998) Power, knowledge and political ecology in the third world. Prog Phys Geogr 22:79–94

Bryant RL, Bailey S (1997) Third world political ecology. Routledge, London

Carney JA (1996) Converting the wetlands, engendering the environment: the intersection of gender with agrarian change in Gambia. In: Peet, R and Watts, M, eds. Liberation ecologies: environment, development and social movements, Routledge, London, 165–187

Dakubo C (2004) Ecosystem approach to community health planning in Ghana. EcoHealth 1: 50–59

Dakubo C (2006). Applying an Ecosystem Approach to Community Health Research in Rural Northern Ghana. Unpublished PhD dissertation. Ottawa

Escobar A (1996) Constructing nature: elements for a poststructural political ecology. In: Peet, R and watts, M (eds) Liberation ecologies: environment, development, and social movements, Routledge, London, 46–68

Fairhead J, Leach M (1995) False forest history, complicit analysis: rethinking some West African environmental narratives. World Dev 23:1023–1035

Forsyth T (2003) Critical political ecology: the politics of environmental science. Routledge, London and New York, NY

Goldman M, Schurman RA (2000) Closing the great divide: new social theory on society and nature. Annu Rev Sociol 26:563–587

Guthman J (1997) Representing crisis: the theory of Himalayan environmental degradation and the project of development in Post-Rana Nepal. Dev Change 28:45–69

Harper J (2004) Breathless in Houston: a political ecology of health approach to understanding environmental health concerns. Med Anthropol 23(4):295–326

Hughes CC, Hunter. JM (1970) Disease and "development" in Africa. Social Sci Med 3(4): 443–493

Joekes S, Leach M, Green C (eds.) (1995). Gender relations and environmental change. IDS Bulletin (Suppl.) 26(1):102 pp

Kalipeni E, Oppong J (1998) The refugee crisis in Africa and implications for health and disease: a political ecology approach. Social Sci Med 46:1637–1653

Leach M (1991) Engendering environments: understanding the West African forest zone. IDS Bulletin. 22:17–24

Leach M (1994) Rainforest relations: gender and resource use among the Mende of Gola. Smithsonian Institution Press, Sierra Leone. Washington, DC

Leach M, Mearns R (1996) The lie of the land: challenging received wisdom on the African environment. James Currey, Oxford

Mackenzie F (1995) Selective silence: a feminist encounter with the environmental discourse in colonial Africa. In: Crush J(ed.) Power of Development, Routledge, New York, NY, pp 100–12, 324 pp.

Mayer JD (1992) Challenges to understanding social patterns of disease: philosophical alternatives to logical positivism. Social Sci Med 35:579–587

Mayer JD (1996) The political ecology of disease as one new focus for medical geography. Prog Human Geogr 20(4):441–456

Mayer JD (2000) Geography, ecology and emerging infectious diseases. Social Sci Med 50: 937–952

Millennium Ecosystem Assessment Series (2005) Ecosystems and human well-being: a framework for assessment; ecosystems and human well-being: Available online at: http://www.millenniumassessment.org/. Accessed 03 May 2010

Morse, S and Stocking M(eds) (1995) People and environment. University College London Press, London

Neumann RP (1996) Dukes, earls and ersatz Edens: Aristocratic nature preservationists in colonial Africa. Environ Plan D14:79–98

Peet R, Watts M (1996) Liberation ecology: Development, sustainability and environment in an age of market triumphalism. In: Peet, R and M Watts (eds.) Liberation ecologies: environment, development, and social movements., Routledge, London, pp 1–45

Peluso NL (1992) Rich forests, poor people: resource control and resistance in Java. University of California Press, Berkeley, CA

Peluso NL (1993) Coercing conservation? The politics of state resource control. Global Environ Change 3:199–217

Rocheleau DE (1995) Gender and biodiversity: a feminist political ecology perspective. IDS Bulletin. 26:9–16

Rocheleau, D, Thomas-Slayter, B, Wangari, E (eds.) (1996) Feminist political ecology: global issues and local experiences. Routledge, London

Schmink M, Wood CH (1987) The 'political ecology' of Amazonia. In: Little, PD and Horowitz, MM, (eds.) Lands at risk in the third world: local-level perspectives, Westview Press, Boulder, CO, pp 38–57

Schroeder RA (1995) Contradictions along the commodity road to environmental stabilization: foresting Gambian gardens. Antipode 27:325–342

Taylor P, Buttel F (1992) How do we know we have global environmental problems? Science and the globalization of environmental discourse. Geoforum 23(3):405–116

Thompson M, Warburton M, Hartley T (1986) Uncertainty on a Himalayan scale: an institutional theory of environmental perception and a strategic framework for the sustainable development of the Himalayas.. Ethnographica, Milton Ash Publications, London

Turshen M (1984) The political ecology of disease in Tanzania. Rutgers University Press, New Brunswick, NJ

Chapter 13
Examining Public Health Concerns from a Critical Perspective

Contents

13.1 Introduction

The new public health discourse proposes a multi-causal and socio-ecological approach to investigating and responding to public health concerns, yet, most public health research, practice, and intervention still focus on individual level characteristics, while relieving the role of broader socio-political and ecological factors. This excessive focus on the individual has led others to criticize the new public health movement as not "walking the talk", and ignoring the social context in which poor health occurs (Poland 1992). The growing complexity of today's health problems, including the rapid emergence of new diseases requires an approach to public health that takes into account the complex ways in which social and political factors interact with biophysical determinants to produce ill health, especially among vulnerable populations. The persistence of spatial and social health disparities in many regions of the world, especially between North-South, urban-rural, Indigenous-Non-Indigenous populations, has caused critics to further question the efficacy of current approaches to health improvement.

Some of these recent criticisms come from critical public health scholars who draw on poststructuralist perspectives, and the politics and sociology of science to interrogate the underlying assumptions and practices of the new public health (Lupton 1995). Like other scientific knowledge claims, critical scholars argue that,

C.Y. Dakubo, *Ecosystems and Human Health*, DOI 10.1007/978-1-4419-0206-1_13,
© Springer Science+Business Media, LLC 2011

the practices and discourses of public health are not value-free, but are socially constructed and can be located. Drawing on Foucault's writing on medicine, the body, and governmentality, critical scholars see public health, and medicine in particular, as sites for the reproduction of power relations, construction of subjective positions and of human embodiment (Foucault 1975; Turner 1988). Through its prescription of "appropriate" and "inappropriate" health behavior, the new public health is seen as a means of organizing and normalizing society (Turner 1994). These prescriptions and norms of the new public health are seen to represent new forms of governance, regulation and social control (Lupton 1994, 1995). However, because public health practice tends to be associated with the "social good" – helping society stay healthy, these normalizing practices are often not recognized as coercive, dominant or controlling. Instead they appeal to widely accepted norms and health practices, and become the commonly accepted body of public health knowledge that sustains society.

Just like environmental problems discussed in the previous chapter, critical public health scholars make use of discourse analysis to examine the discursive practices through which public health problems are constructed, explained and intervened upon. It also interrogates the taken-for-granted assumptions underlying public health knowledge and practices and examines how subject positions are constructed. Critical public health broadens its analysis of the determinants of ill health to encompass socio-political, ecological, and historical antecedents, and refuses to accept explanations of poor health at face value.

13.2 Public Health as Discourse

A *discourse* is broadly defined as a truth regime that relates to specific social phenomena or practice and is mostly expressed through texts and the use of language (Hajer 1995). Specific discourses share common understandings of the phenomenon in question, the causes associated with it and the appropriate solutions (Adjer et al. 2001; Aviles 2001). A discourse that is pervasive and is held by many people in society as the main understanding of a particular phenomena is described as a dominant discourse, as seen in public health or medicine. The production of such dominant discourses is linked with the exercise of power, as power is required to legitimize and sustain it.

Within the health field, many health problems are knowable through medical science. Health professionals are seen as the legitimate experts who can name, explain, and solve a particular health problem. They also are able to define what values, beliefs, behaviours, and practices others should hold (Lupton 1998). The privilege of the medical professional to name, define and solve another's problem wields power in the professional. In order to sustain this power, the knowledge claim must be validated and seen as legitimate. The validation, maintenance and circulation of medical knowledge claims take various forms including boundary setting, and the use of specialized language, as discussed in Chapter 11 (Aviles 2001; Lee and Garvin 2003).

Boundary setting usually takes place through the identification of who is allowed to speak, from what viewpoint, and what is allowed and not allowed (Aviles 2001). It is through boundary setting that "lay", and "indigenous" perspectives get silenced or excluded as valid ways of knowing. By specifying what can and cannot be said, boundary setting is able to specify what behaviours and practices are appropriate and which are not. For example, in many developing countries, the health practices and beliefs of local people are sometimes considered "out-moded", "primitive" or "health deteriorating." Local people's understanding of health or disease is seen to be inferior to scientific understanding. However, by virtue of the fact that there are alternate views of health and illness, this means that health and illness are social constructions, and the views of the health professional or scientist should not be given special privilege (Eyles 1993). Lay experiences, beliefs and perceptions are important in gaining contextual understanding of health and health problems.

In addition, validation of medical knowledge systems, is achieved through the use of expert "scientific" language that can be understood by a few. Medical professions use labels such as "sick" "pre-disposed" or "at risk group" to describe various subject positions. These labels are constructed based on abnormalities from the preconceived "normal" or "healthy". Having been constructed as "abnormal", the medical expert is then justified to intervene and propose some appropriate health interventions. These interventions often appeal to the social good, and to community health and well-being, while denying and concealing any political, social or economic interests. Because many public health problems tend to be diagnosed from a biomedical perspective, and solely by the health professional, underlying micro-struggles and socio-political factors fail to be captured succinctly, therefore rendering proposed interventions ineffective.

Also, because of the privileging of medical/health knowledge over lay perspectives, there is the tendency to assume that lay people are "empty vessels" or passive consumers of this knowledge system, and would readily comply with the recommended solutions. However, lay people are able to resist dominant discourses by reconstructing them in light of their own social realities and traditional knowledge systems, and deciding whether or not to comply and implement the proposed interventions. They also are able to resist dominant knowledge systems through avoidance behaviours and non-participation in health programs. These resistance activities illustrate the agency and adaptive powers of lay people, further illustrating that power does not necessarily concentrate on the scientific expert.

13.3 Negotiating Definitions of Health and Ill Health

In many instances, public health researchers and professionals have defined "good health" and "poor health" for individuals and communities. They have also identified and named problems for them. However, in many instances, the definitions and interpretations of these health problems often differ from those of lay people (Dakubo 2004). For example, in many developing countries and Indigenous communities, people tend to conceive health from a holistic and ecological perspective, seeking a balance among mental, spiritual, emotional, and physical health. Most health professionals, on the other hand, tend to conceptualize health from a biomedical perspective. The biomedical model of health is the predominant model of health across various societies and views disease as a deviation from normal biological functioning, which needs to be fixed therapeutically (Gordon 1988). However, critics argue that such notions of disease and health objectifies the human body as a machine to be "fixed" by medical experts, and idealizes the "normal" human body as being in a perfect state of health. Any human body that comes short of the normal standard of an idealized body is not healthy and is need of repair (Lee and Garvin 2003; Lupton 1995). Such objectification of the human body denies the individual's context and life situation.

In the case of communities, the structural imperatives that affect, produce, and reproduce people's health and well-being are ignored. Community members need to express their own view of health, what they consider to be their health problems and participate in finding solutions them. As illustrated in the Ghanaian case study discussed in chapter six of this book, there were major differences between local people's conception of health and those of community health workers. While most community health professionals conceptualized health from a biomedical perspective, many participants explained health predominantly from psychosocial and ecological perspective. Health was described as the ability to perform societal roles, access basic services, meet personal needs and social obligations, and cope with everyday life circumstances. Participants' conceptualization of health were not static but were expressed in ways that captured the complex, socialized and cultural dimensions of their lived experiences. These conceptions are in line with Kelly et al. (1993) assertions that health has no stable meaning, but change with the uniqueness of individual situations. Such holistic conceptualizations of health requires that public health research and practice broaden its horizon and become more accepting of such notions of health as authentic, as well as incorporate them in health improvement strategies.

Similarly, critical scholars are concerned about how the identities of "sick" and "healthy" are constructed. They argue that people are being constituted in and through discourses and social practices, and so the subject positions of "sick" and "healthy" are discursive categories created through the use of binary logic (Fox 1994). We divide the world into two worldviews of "healthiness" and "sickness", and start from the premise that all healthy individuals must exhibit certain attributes; and all those who fail to exhibit these attributes are not healthy. Critical scholars seek to illustrate that the experiences of "health" and "sickness" are not fixed or static, but are constantly in flux, as well as being negotiated.

13.4 Negotiating the Determinants of Health Problems

Besides the focus on individual level factors, such as inappropriate health behaviours and practices, other determinants of poor health tend to focus on a number of issues, including poverty, limited access to health care resources and services, and exposure to environmental hazards, especially in the developing country context. Local people's traditional values, cultural practices, belief systems, and customs are sometimes seen as obstacles to promoting health and well-being. These practices are perceived as self-destructive, irrational, and stand in the way of modern medicine (Farmer 2001; Harper 2004). Similarly, environmental-related diseases are attributed to inappropriate interaction with the environment or irrational use of environmental resources. Yet, these causes alone are not sufficient to explain the persistence of poor health outcomes and health disparities that characterize our society.

For example, the explanations of poor health outcomes that focus on culture and peoples' belief systems, relieve the role of structural inequalities and power imbalances that perpetuate uneven patterns of health. They also fail to reckon the discursive means through which local peoples' identities and health practices are constructed to suit the ideological, political and material interests of dominant discourses. By focusing on cultural practices, local people are constructed as homogenous by virtue of their shared culture. This homogenous lens is sometimes reflected in health promotion strategies that take the form of "community intervention packages," with little consideration for differential health experiences. Similarly, health explanations that focus on the irrational use of environmental resources fail to examine the underlying socio-political factors that lead to such interactions as discussed by the political ecology analytical framework. Political ecology contributes to this understanding by illuminating how environmental change is constituted through power imbalances among various actors, and how ill health is a product of that politicized environment (Bryant 1998; Mayer 1996).

Within the public health domain, then, critical perspectives require that explanations for the causality of ill health be broadened beyond observable and verifiable phenomena, to include an examination of the socio-political, economic and historical factors that underlie the processes and structures causing ill health and health disparities.

13.5 Historicizing Health Problems

Examining health problems from an historical perspective allows for a better understanding of the social and spatial patterning of health, especially in regions with a colonial legacy. As such, many developing country nationals and Indigenous populations health experiences should be examined from an historical perspective. For example, in the Ghanaian case study, a good understanding of the social and spatial disparities that characterize northern and southern Ghana, and rural and urban regions in the country was best explained through the colonial legacy and the health and public policies that were in place at the time. In Ghana, the main objective of colonialism was to establish and strengthen favourable sociopolitical conditions for imperialist penetration and exploitation. This involved the extraction of natural resources and other primary goods for capital investment (Aidoo 1982). Most infrastructure, such as schools, hospitals, roadways and economic projects were situated in port towns and administrative centres, which were located in the Southern portion of the country (Arthur 1991). The northern half of the country was considered as a "labour-reserve" to serve the mineral exploitation in southern Ghana (Songsore 1989).

Colonial health policies were primarily concerned with preserving the health of European masters, understanding the aetiology of tropical diseases and developing technologies to cure them (Aidoo 1982). As such, disease and parasitic models dominated colonial medical thinking about Africa's health problems. There was emphasis on finding cure for diseases like malaria, yellow fever and schistosomiasis to allow for the expansion of European capitalist activities and colonization of the tropics (Farley 1991, cited in Randall 1998). With this system of health, colonial experts made most of the decisions about health, health care and its distribution (Randall 1998). Indigenous views and health needs were rarely considered as important. Also, the broader determinants of health received little attention. Colonial medical authorities viewed broad-based efforts to deal with the underlying social and economic determinants of ill health as impractical, expensive, and unnecessary (*ibid*). At best, separate housing was developed to protect the health of expatriates. Bungalows with modern water supply and sanitary disposal systems were built for top government officials and merchants of mining companies (Twumasi 1981). Thus, the provision of health services was selective. Although Ghana has since gained independence, this colonial model of health still exists and many state-of-the-art health facilities are still located in the South. There are also still significant health disparities between the North and the South, as well as rural and urban areas.

13.6 Conclusion

This chapter discusses the application of critical perspectives to public health, drawing attention to how public health is deployed as a discourse, and the various ways in which it gains legitimacy and maintains its dominance. Critical public health scholars interrogate public health knowledge claims and interventions, and caution against the uncritical adoption of health concepts and explanations without analyzing them for their social and political baggage. They encourage that health problems be examined from a broader perspectives and be linked to rigorous political economic analytical frameworks.

The adoption of a critical lens also requires that we examine health problems from an historical perspective to allow for a better understanding and interpretation of existing health conditions and patterns. In the case of many African countries and Indigenous communities, it is important to situate their respective health experiences within the context of their colonial legacies, and to examine how colonial political economy and health policies shaped and continues to shape existing health outcomes.

References

Adger WN, Benjaminsen TA, Brown K, Svarstad H (2001) Advancing a political ecology of global environmental discourses. Dev Change 32:681–715

Aidoo TA (1982) Rural health under colonialism and neocolonialism: a survey of the Ghanaian experience. Int J Health Serv 12:637–657

Arthur JA (1991) Interregional migration of labour in Ghana, West Africa: Determinants, consequences and policy. Rev Black Pol Economy 20:89–114

Aviles LA (2001) Epidemiology as discourse: the politics of development institutions in the Epidemiological Profile of El Salvador. J Epidemiol Commun Health 55:164–171

Bryant RL (1998) Power, knowledge and political ecology in the third world. Prog Phys Geogr 22:79–94

Dakubo C (2004) Ecosystem approach to community health planning in Ghana. EcoHealth 1: 50–59

Eyles J (1993) Feminist and interpretive method: how different? Can Geographer 37:50–52

Farmer P (2001) Infections and inequalities: the modern plagues. Updated Edition University of California Press, Berkeley

Foucault M (1975) The birth of a clinic: the archaeology of medical perception. Vintage Books, New York, NY

Fox NJ (1994) Postmodernism, sociology and health. University of Toronto Press, Toronto, ON

Gordon DR (1988) Tenacious assumptions in Western medicine. In: Lock M and Gordon D (eds) Biomedicine examined. Kluwer, Dordrecht, pp 19–56

Hajer MA (1995) The politics of environmental discourse: ecological modernization and the policy process. Clarendon, Oxford

Harper J (2004) Breathless in Houston: a political ecology of health approach to understanding environmental health concerns. Med Anthropol 23(4):295–326

Kelly MP, Davies JK, Charlton BG (1993) Healthy cities: a modern problem of a postmodern solution? In: Kelly MP and Davies JK (eds) Healthy cities: research and practice. Routledge, London, pp 159–168

Lee RG, Garvin T (2003) Moving from information transfer to information exchange in health and health care. Soc Sci Med 56:449–464

Lupton D (1994) Medicine as culture. Sage, London

Lupton D (1995) The imperative of health. Public health and the regulated body. Sage, London

Lupton D (1998) The emotional self: a socio-cultural exploration. Sage, London

Mayer JD (1996) The political ecology of disease as one new focus for medical geography. Prog Hum Geogr 20(4):441–456

Poland BD (1992) 'Learning to 'walk our talk': the implications of sociological theory for research methodologies in health promotion. Can J Public Health 83(Suppl 2):S31–S46

Randall P (1998) Health care systems in Africa: Patterns and Prospects. Report from the workshop, Health Systems and health care: Patterns and Perspectives, 27–29 April 1998. The North-South Co-ordination Group. University of Copenhagen and The ENRCA Health Network

Songsore J (1989) The spatial impress and dynamics of underdevelopment in Ghana. In: Swindell K, Baba JM, Mortimore MJ (eds) Inequality and development: case studies from the third world. Macmillan, London

Turner BS (1988) Medical power and social knowledge. Sage, London

Turner BS (1994) Theoretical developments in the sociology of the body. Aust Cult Hist 13:13–30

Twumasi P (1981) Colonialism and international health: a study in social change in Ghana. Soc Sci Med [Med Anthropol] 15B(2):147–151

Chapter 14
Towards a Critical Approach to Ecohealth Research and Practice

Contents

14.1 Introduction

Central to the field of ecohealth is the notion that human beings are integral to nature, yet analysis of the social and political dynamics that produce environmental degradation, resource depletion, and consequently ill health remains undertheorized. Most of the literature adopts uncritical notions of environment and health, and accept a priori notions of ecological causality and change, without questioning the political contexts within which such explanations emerge and become relevant. The concern for rigorous investigation of human-environment interactions and how these produce various vulnerabilities and ill health is at the core of critical theoretical developments, including critical political ecology and critical public health.

The past three chapters illustrate the application of critical theoretical perspectives to environment and public health disciplines. As an emerging, integrated and under-theorized field, *Ecohealth* stands to benefit from such new theoretical developments. The goal of this chapter, then, is to outline an analytical framework for

C.Y. Dakubo, *Ecosystems and Human Health*, DOI 10.1007/978-1-4419-0206-1_14,
© Springer Science+Business Media, LLC 2011

ecohealth that takes into account these new theoretical developments when investigating problems at the interface of health and environment. It will do so by integrating two theoretical perspectives; *critical political ecology* and *critical public health* to develop a *critical ecohealth framework*. *Critical ecohealth* draws its critical stance from a variety of fields, including political ecology, sociology of science, poststructuralism, postcolonial development, and feminist theories.

There are a number of key issues that are central to this new framework. First, it is important to acknowledge that ecohealth is a discipline that produces knowledge claims related to problems at the interface of health and environment. Like other disciplines, the knowledge claims produced by ecohealth are socially constructed and not value-free. The production of ecohealth knowledge claims should be situated within the socio-political and historical contexts within which they are constructed. Ecohealth knowledge claims should be interrogated for their construction of an ecohealth problem, either from a health or environmental perspective, how they explain the causal factors responsible for the problem, how they construct various subject positions, and the type of interventions proposed. In addition it is important to examine the ways in which ecohealth, as a body of knowledge, is produced and circulated, and the processes through which it claims authority and legitimacy. The adoption of such a reflexive stance allows the ecohealth researcher or practitioner to understand knowledge as situated, partial, and is continuously being negotiated.

Particular attention has to be paid to how ecohealth problems are constructed, by who and for who? Critical ecohealth seeks to avoid the uncritical adoption of a priori notions of ecosystem degradation and health problems, without evaluating such constructions within the broader sociopolitical and historical contexts in which they occur. It cautions against adopting simplistic explanations for ecosystem degradation and poor health status, without revaluating these within power structures that characterize society.

Thirdly, critical ecohealth recognizes that the costs and benefits of ecosystem change and degradation are not evenly distributed or equally experienced, and so attention must be paid to particularities of the victims. The analysis of these experiences should be contextualized based on the myriad, changing, and conflicting identities of those affected. These issues are elaborated upon along the themes of the ecohealth approach.

14.2 Identifying Environmental Problems from a Critical Perspective

One entry point for an ecohealth research project is to identify an environmental condition that is suspected to have an adverse impact on human health such as a degraded ecosystem, a polluted water body, a mining site, or a slum in an urban community. Alternatively, ecohealth research could start with a human health condition that is suspected to be influenced by a particular ecosystem condition. In both instances, many researchers commence their research by adopting the pre-existing notions of environmental degradation or the health problem to be investigated, without critically evaluating how the problem came to be labeled as a degraded environment, by who, and for who? What procedures were used to construct the ecosystem as degraded? What role did those affected play in constructing the ecosystem as degraded?

The concern is that by adopting pre-defined or pre-existing concepts of ecosystem degradation, usually informed by scientific indicators of a healthy ecosystem, we could be misinterpreting a natural phenomena or a deliberate ecosystem modification by local people. We also could falsely be adopting a label that was framed based on the social and political interests of the framers, and has nothing to do with the actual biophysical condition of the ecosystem. Alternatively, the condition of the ecosystem could be a result of an historical experience that is not readily apparent to the researcher. The adoption of a priori notions of ecosystem degradation could lead researchers to make a wrongful diagnosis of the problem, develop interventions that do not address the problem, or use findings to inform policy development that further constrain people's access to ecological resources. In the case of an eco-health research project, this could lead to making wrong associations between the ecosystem condition and a particular health problem.

How then can we go about identifying or naming an environmental problem that takes into account social, political, historical and ecological realities. Instead of adopting a pre-existing definition of an environmental problem, it is important that the team of transdisciplinary researchers and the people affected collaboratively name the problem. The joint identification of the problem will allow for the integration of scientific and local perspectives, while making transparent that the problem was not constructed to benefit the political, economic and academic interests of the researchers. Local perspectives will also fill in the gaps on historical and cultural factors, ensuring that the problem is correctly interpreted. Critical ecohealth therefore, seeks to produce a hybridized understanding of ecosystem degradation that is both socially and biophysically relevant, and a health problem that incorporates local understanding. When an ecohealth problem is mutually identified, the chances that interventions emanating from such democratized problem identification will succeed are higher, compared to those based solely on the perspective of the "expert" scientist.

In addition, the joint identification of the problem allows researchers to tailor interventions to the particular needs of the people being impacted. Critical ecohealth cautions against framing ecohealth problems from a "global" perspective, as this fails to recognize the unique and specific challenges faced by various population groups. The framing of problems from a global perspective also implies an uncritical adoption of dominant constructions of environmental problems as "global" in scope and "crisis" in nature. By super-imposing a global lens on local environmental problems, we fail to better understand exactly how a small community is being affected by an environmental problem. Who is affected and how? For example, while climate change is a global phenomena, it is still important to ask the community of 50, how they think they will be impacted by climate change, and tailor adaptation measures to their specific circumstances. It is also important to realize that people's circumstances are not static, but are constantly changing, and any proposed interventions must take this into account and be adaptable rather than fixed.

14.3 Explaining Environmental Problems from a Critical Perspective

After having identified and named the environmental problem at hand, the next phase of the research attempts to understand and explain the driving forces behind the occurrence of the problem. The most common explanations for environmental degradation tend to focus on consumerism capitalist expansion in the North, abject poverty, rapid population growth, and poor land use practices in the South; and in transitioning countries like India and China, inappropriate or mal-development activities such as construction large scale dams, and intensive agricultural production, systems. Most of these factors are aligned with the dominant global environmental management discourse (Adger et al. 2001) described in Chapter 12, in which the ordinary citizen, local people and farmers are seen as the perpetrators of ecosystem degradation, as well as the victims of their own activities. The populist discourse, on the other hand, sees local people and their environments as victims of capitalist activities and the forces of globalization. Critical ecohealth suggests that it is insufficient to try to explain ecosystem degradation based of these causes alone, without locating their emergence in the context of unequal power structures and the organization of broader society. For example, drawing on political ecology, most poor land use practices do not emerge in a vacuum, but instead are the result of the unequal power relations surrounding the use of, and control of land-based resources. Powerful actors including the state, industries, and private sector, may control large portions of land thereby forcing weaker actors to encroach onto marginal lands. With very limited options, these weaker actors then make use intensive land-use practices so as to maximize productivity from marginal lands. So, the superficial explanation of poor land-use practices does not only fails to capture the underlying problem, but also unduly blames the victim.

In addition, ecosystem degradation at the local level must be seen as being shaped by broader macro-level structures, such as state, national and international policies and regulations (Blaikie 1995; Bryant 1997). The causal factors shaping ecosystem degradation must be broadened to encompass social, political and economic factors at varying scales, and how the interaction of these factors leads to ecological change and adversely impacts health. Nature and society are co-constitutive, and the physical explanations of ecosystem degradation must not be seen as independent of the social context in which they occur, nor are they ever politically neutral (Freundenberg et al. 1995). Similarly, the causal basis for the onset of ecosystem-mediated health problems (e.g. schistomiasis) should be examined beyond the biophysical environment (e.g. dam construction), and instead evaluated based on the political economic factors that produced the so-called "diseased environment".

It also is important to historicize the occurrence of environmental problems and draw on multiple perspectives to help researchers develop a comprehensive understanding of the causal basis of environmental problems. These varied perspectives and experiences need to be examined or interpreted through people's unique identities and changing circumstances.

The goal of critical ecohealth, then, is to avoid the adoption of simplistic explanations of ecosystem degradation, and to develop a socio-politically, yet biophysically grounded understanding of the causal factors shaping ecosystem degradation. It calls for broadening the explanations for the causal basis of ecosystem degradation to incorporate an analysis of the uneven distribution of power in society and how this constraints peoples access to ecological resources.

14.4 Identifying Health Problems from a Critical Perspective

As mentioned above, one entry point for ecohealth research is to identify a health problem that is suspected to be influenced by degraded or poor environmental conditions, such as in malaria or dengue associated with deforestation. Just like environmental problems, it is important to incorporate the voices of those affected in naming the particular health problem to be investigated. While the ecohealth researcher might identify the health condition by name (e.g. malaria), based on medical signs and symptoms, those affected, especially in developing countries and Indigenous communities, may not necessarily describe their health problems from a disease perspective. Studies suggest that people from these communities tend to describe their health experiences from a broad, and holistic perspective (Dakubo 2004). So by referring to the health condition by name, the ecohealth researcher narrows their investigation and proposes interventions that respond solely that particular health problem, while ignoring other elements that could have contributed to promoting overall individual and community well-being. The opportunity for beneficiaries to participate in naming their health problems, allows for a better understanding of how they conceptualize health and ill health, as well as what they perceive to be the underlying causes of those health problems. In the case of an ecohealth research, such collaborative identification allows the researcher to gain a better understanding of the emergence of the environmental health problem, thus, allowing for the development of appropriate interventions that not only treats the health condition, but also seeks to respond to the underlying factors through sectoral collaboration. Critical ecohealth therefore cautions against the uncritical adoption of pre-defined notions of health and health problems, given that health and illness are social constructions and need to be contextualized.

Like constructions of environmental problems, certain health problems are constructed from a global and "pandemic" perspective, triggering the need for the development of universal interventions. However, as noted throughout this book, constructions of either environment or health problems should not be taken as a neutral process, but must be interrogated for their associated baggage. For example, in the case of HIV/AIDS, how do we come to understand HIV/AIDS as so pervasive in Sub-Saharan Africa? Why does malaria not command similar attention and publicity, given that the burden of malaria is higher than HIV/AIDS? Who constructs and prioritises these health problems?

HIV/AIDS is one of those health problems that assumes figures that cannot be substantiated, especially in Africa. As part of a research project in a rural community in Ghana, West Africa, I participated in an outpatient health education session on HIV/AIDS awareness and prevention. Following the session, I asked the community nurse whether HIV/AIDS was a major health problem in the community? Her response was no, not a single case had ever been reported in the community. My follow-up question then was, why was an education session held on HIV/AIDS and not on malaria prevention, given that over half of the outpatients who participated in the education session had come to the clinic with reported cases of malaria? The response was that they had received directives from the Ministry of Health to intensify HIV/AIDS education and awareness in their communities, and resources in the form of vehicles had been provided for this exercise.

This example is indicative of the implications of the global construction of health problems and prescription of universal solutions. HIV/AIDS has become synonymous with Africa. Because of its widespread construction and publicity, many communities are forced to implement HIV/AIDS campaigns, dedicate enormous time and resources to it, while waiting to encounter their first case. Critical ecohealth does not refute the fact that HIV/AIDS is of major concern in certain regions of Africa, what it is concerned about is the undue emphasis and concentration of health care resources on HIV/AIDS in regions with little or no incidence, when such resources could have been used to help address an equally, if not more important health problem such as malaria. Such "packaged" universal interventions do not only fail to respond to health problems in Africa, but they also contribute to increasing the mortality and morbidity rates of diseases like malaria because of diverted health care resources.

Critical ecohealth argues that the construction of HIV/AIDS, and other "global" health problems should always be interrogated for their political and social framings. Many health problems in Africa are externally determined and prioritized by global centres of expertise like the World Health Organization, hence the extent to which health care is organized to respond to "actual" problems is minimized. In addition, many external institutions and non-governmental organizations interested in HIV/AIDS donate medical resources and equipment with conditionalities such as the institution of educational programs which must be heeded. Critical ecohealth posits that, rather than work with externally prioritized health problems, it is important to collaborate with communities and regions to identify and prioritize their own health problems. It is by working directly with communities and on community-identified health concerns, that targeted and culturally appropriate interventions can be developed. The goal of critical ecohealth then is to integrate local perceptions of the health problem under investigation, with scientific understandings from the transdisciplinary research team, so as to develop a transparent yet accountable form of knowledge that is grounded in people's health experiences and needs.

14.5 Explaining Health Problems from a Critical Perspective

The causal basis of many ecologically-induced health problems tend to focus on people's inappropriate interaction with the elements of the biophysical environment that exposes them to disease vectors and pathogens. For example, people come into contact with disease vectors, through various land use activities such as deforestation, irrigation, clearance of virgin land, and agriculture. Similarly, many water-borne diseases such as diarrhea focus on the use of contaminated water or engagement in unhygienic practices. Many respiratory diseases in developing countries tend to be attributed to indoor air pollution, resulting from the use of cow dung and crop residues as sources of fuel. Traditional and cultural belief systems and values are sometimes seen as factors that inhibit the adoption of good health practices.

These causal factors are superficial and do not sufficiently analyse the social and political forces that mediate society-environment interactions and produce diseased environments that negatively affect human health. Critical ecohealth sees ecologically-mediated diseases and ill health as resulting partially from the unequal power relations surrounding the use of, access to, and control of ecosystem resources and services. In developing countries, in particular, unequal power relations force weaker actors to engage in land use practices that further augment their vulnerabilities and exposure to pathogens and disease vectors. Land use policies and legislations have also constrained access to nutritious wild food and game in forests and other ecosystems. Critical ecohealth seeks to illustrate that environmental health problems are shaped by political, economic factors' that must be examined within the broader frameworks of power struggles surrounding the use and distribution of environmental benefits and costs (pollution). These uneven power relations provide a means of understanding how the complex interplay of socio-ecological and political factors shape the spatial and social patterning of various' health outcomes. Critical ecohealth seeks to broaden the examination of the underlying causes of ecosystem-related health problems beyond simplistic structural explanations and exposure to pathogens, to include the social organization and power dynamics characterizing human-environment relations.

In addition, it is important to examine how micro-level factors interact with macro-level forces to shape people-environment interactions that may adversely impact health. For example, how do state policies, globalization and unfair trade agreements influence people-environment relations? By casting the analysis of causal factors within interacting forces at various temporal dimensions and scales, the researcher is able to examine how patterns of health and disease are linked to the power of local elites, to state practices, and to the global policies and processes of capital accumulation.

Finally, it is important to locate explanations of health problems and their social patterning within the particular history of a region. For example, the spatial and social patterning of health between Northern and Southern Ghana, and between rural and urban areas can mostly be understood through colonial policies. Hence, in the context of many African countries, it is important to examine how colonial environment and health policies shaped people-environment relations, and whether or not some of these policies are still in effect. In the context of Africa, macro-economic policies by the World Bank and International Monetary Fund, such as structural adjustment programs played a major role in shaping people's health outcomes, as well as people's interaction with the biophysical environment.

Critical ecohealth seeks to move beyond the biomedical focus on pathogens and disease vectors to incorporate broader social, political, and historical processes at the local, regional and national levels into the causal analysis of poor health out-comes. The goal is produce causal explanations that will allow for the development of focused and targeted, yet people-centred interventions that will promote both ecosystem and human health.

14.6 Contextualizing Environment and Health Experiences

Ecohealth recognizes that the benefits and costs of ecosystem activities are unevenly distributed, and the experiences and coping abilities associated with these costs and benefits vary from person to person. Thus, ecohealth research makes use of procedures that take this social variation into account, ensuring that it develops interventions that are socially equitable. Some of these research procedures lead as to conduct research based on various axes of difference including gender, age, socioeconomic status, among others.

Critical scholars caution that by focusing on specific groups, we make assumptions that there is a particular experience associated with that group, and through the use of appropriate research methodologies we are able to capture and describe those experiences. As critical theorists point out, there are no fixed experience, as our accounts of reality, and experiences are mediated by our multiple, varying, and conflicting positions. These multiple and changing positionalities produce many ways of interpreting, understanding, and viewing the social world. As such, there are multiple realities that need to be contextualized based on our changing positions (Butler 1992; Fox 1994).

Critical scholars also caution that in our attempt to pin down particular experiences, we run the risk of essentializing the category (i.e. the group or individual) as natural, with the tendency to make use of totalizing or universalizing discourses to describe these varied and complex experiences. For example, Whiteford and Manderson (2000) argue that Southern women's health and health experiences tend to be framed from a universal perspective and presented as though women have similar, or shared health concerns. Such totalizing representation of women's health needs and experiences have failed to address important psychosocial health concerns of women, especially in developing countries, as these do not usually fit within the mainstream women's health agenda, such as maternal and child health (Dakubo 2004). Critical scholars argue that the use of totalizing discourses smoothens over varied experiences, and creates an illusion of homogeneity and coherence. There is also the risk of interpreting people's life experiences out of an historical context, thus creating a false appearance of timelessness (Abu-Lughod 1993).

Critical ecohealth rejects totalizing discourses, and emphasizes the need to pay attention to specificity, and to contextualize people's environment and health experiences in relation to their respective daily interactions with their social, physical and economic environments. Rather than look for a specific environment or health experience, we should look for multiple experiences, and how these multiple experiences are being produced, reproduced and negotiated. In order to capture these multiple and varying experiences, research methodologies will have to focus on the particularities of people's lives, making use of qualitative and ethnographic methodologies such as life histories, narratives, and in-depth interviews.

Critical ecohealth also seeks to examine how subject positions are constructed in ecohealth, especially in relation to gender. "Gender" is a socially constructed category that refers to culturally produced roles, behaviors, and identities (Butler 1990). However, accordingly to Whittle and Inhorn (2001), some health research replace "sex" (which is the biological category) with "gender" in their analysis. The replacement of gender with sex does not allow for an adequate examination of the implications of "gender" roles and relations on health. For example, we are not able to adequately examine how women's daily lives are influenced by gender norms and expectations of being a woman ("the good wife") or a man; how gendered relationships among and between men and women influence health status; how gender inequality, perpetuated by institutional structures, affects women's lives; or how the effects of gender are cross-cut by other identities such as race/ethnicity, age, and class (Whittle and Inhorn 2001:155).

The focus on gender, and especially women's health, has drawn criticisms from critical feminists who argue that undue emphasis on gendered health outcomes, may result in the unwanted effect of representing "gender as an essential, irreducible part of identity" (Frug 1992:36). They argue that the search for a "universal women's experience" may result in the oversimplification of women's health experiences. The focus on women rather than gender, could result in us ignoring men's health as important (Annandale and Clark 1996). By ignoring men and treating women as distinct from men's, we are inevitably constructing women's health as "poor" and in need of attention against an implicit assumption that the male body is healthy (ibid). The use of such binary logic does not only privilege men as "healthy" and women's health as "unhealthy", but also undermines our ability to understand women's health as shaped by other axes of difference. Following this then, feminist poststructuralists require that we dislodge the opposition between men and women, avoid seeking commonalities, and instead show the complexities of both men and women's lives as embedded in ethnicity, class, religion, and place (Barnes 1982).

However, some feminists are against using an approach that does not feature women's issues prominently, as this could diffuse feminist politics. They argue that the focus on women's issues is one way to draw attention to the challenges and oppressive structures of women's lives, and so any attempt to minimize this focus may end up augmenting the inequities that have served as the basis for many feminist struggles. According to Smart (1990), though, the goal of poststructuralism is not to eradicate the politics of gender, but to reject totalizing discourses and their ability to provide "truthful" answers to problems. To overcome this dilemma, Spivak and Ronney (1989) and others suggest the "strategic" use of theory. In such an approach, women's lived experiences, including their health and environment experiences, could sometimes be improved by deploying "women" as a category and sometimes by emphasizing the plurality and multiplicity of experience which may intersect gender (Butler 1990).

Within the context of ecohealth then, the strategic use of gender is very important. For example, in many developing countries, men and women tend to interact with the biophysical environment in vary different ways, which leads to the acquisition of gendered knowledge patterns (Mackenzie 1995; Rocheleau 1995) and exposure to different health risks. Similarly, the household is a site of power struggle, where gender roles and relations may influence men and women's relationship with the environment differently and expose them to different health risks. In these circumstances, there may be opportunities where focusing on women or men may provide a better insight to gender-disaggregated knowledges, environmental or occupational health risks. In the same vain, it may also be interesting to examine how men and women's environment and health experiences are shaped by their multiple identities, of which gender is just one.

In sum then, critical ecohealth hopes to avoid the construction of subject positions, including individuals, groups and communities, as though they had pre-existing identities. The goal is to contextualize environment and health experience based on the multiple and conflicting identities of people, as well as, strategically deploying various axes of difference.

14.7 Developing Mutually Acceptable Ecohealth Interventions

Ecohealth research projects seek to develop interventions that simultaneously promote human health and ecosystem health. This is based on the notion that improving health through better ecosystem management is a cost-effective approach to disease prevention and health promotion, as opposed to accessing scarce and expensive medical services. However, the production of ecohealth interventions are closely linked to the production of ecohealth knowledge, both of which are inherently bound up with power relations. From this perspective then, the production of "appropriate" ecohealth interventions are forms of knowledge claims that are not value-free, but are bound up with the exercise of power. Ecohealth interventions specify what environment and health activities to implement, so as to improve human health and ecosystem health. In order to maintain this authority and legitimacy, ecohealth draws on scientific evidence, sometimes backed by local perspectives, that appeal to the common good by seeking to promote human health and ecosystem health. By appealing to notions of "common good" and "scientifically proven interventions", ecohealth researchers and practitioners are able to maintain authority, be seen as legitimate, while concealing or rendering natural any political and economic interests (Escobar 1996; Guthman 1997; Schmink and Wood 1997).

Critical ecohealth seeks to avoid the uncritical adoption of environment and health interventions as benevolent, instead suggesting that interventions be examined for their social and political contexts. Through this, interventions that may augment social inequities and vulnerabilities can be avoided. It is also important to realize that inappropriate interventions can always be resisted through forms of resistance and other non-compliance measures.

Critical ecohealth also seeks to work with community members and all relevant stakeholders in ways that will enhance the acceptability of ecohealth interventions. The first step in ensuring that interventions are mutually acceptable to both community members and the transdisciplinary team is to actively involve beneficiaries in all stages of the research process, as called for by participatory action research. The involvement of community members as co-researchers enhances the validity of problem identification, explanation and resolution. The likelihood that interventions that emanate from such engagement will be accepted is high. Similarly, interventions that are developed without taking sufficient steps to contextualize people's experiences may end up making use of universal solutions that fail to respond to specific needs.

Local people have unique insights and understandings of environment and health issues that are not readily apparent to researchers. Hence it is encouraged that interventions incorporate these insights, while being cognizant that not all forms of local knowledge are accurate, complete, or current. The use of hybrid research and multiple data sources will complement local knowledge to provide a better understanding of the issue under investigation. Also, ecohealth research sometimes addresses complex, and sometimes technical environment and health concerns that are not suitable to local understanding. For example, examining or explaining the pathways through which mercury or lead poisoning, and other agricultural chemicals could affect the nervous system is probably too complex for local people to comprehend. In such situations, critical ecohealth recommends that specialists or members of the transdisciplinary research team work with local people to arrive at basic understanding of the issue, so as to enhance the acceptability of proposed interventions. The integration of expert and lay perspectives in the development of ecohealth interventions ensures that it is neither a top-down, nor a completely bottom-up intervention strategy, but one that is mutually acceptable and responds to community needs.

Finally, it is important to realize that not all proposed interventions will be accepted by the beneficiaries as appropriate the "truth", as these are usually re-evaluated in light of their own health and environment experiences, before making a conscious and informed decision as to whether or not to adopt or implement them. If interventions were developed unilaterally and do not reflect their needs or voices, they are able to resist or challenge these interventions in ways that are not readily apparent to other environment and health professionals. This culture of resistance takes various forms in both the public health and the environment sector. For example in the public health sector, this could take the form of non-adoption of health promotion strategies, non-participation in health education programs, and the use of other non-biomedical forms of healing. In the environment sector, resistance could take the form of "inappropriate" land use practices, illegal poaching or tree felling, bushfires, and arson. Ecological resistance activities are inscribed in the environment, in ways that provide another way of reading the environmental "text" of weaker actors and the power relations in which they are embedded (Bryant and Bailey 1997).

However, because these resistance activities may be misrepresented by "experts" as poor land use practices, it is difficult for practitioners to see these activities as forms of resistance. Critical ecohealth seeks to draw attention to alternative ways of interpreting poor land use practices and the so-called "inappropriate" health behaviours as forms of resisting dominant ideologies and interventions that do not serve communities needs. Participatory approaches should be used to ensure that interventions are democratically developed, socio-politically informed, and mutually acceptable.

14.8 Approaching Transdisciplinarity and Participation from a Critical Perspective

Methodologically, the ecohealth approach calls for the use on transdisciplinary procedures that bring together a team of professionals from the natural, health and social sciences. This team of experts works with local people and other relevant stakeholders to plan and conduct research, including data gathering, analysis, and implementation of appropriate solutions. Ecohealth research also calls for the use of participatory procedures that actively involve community members in the research process.

There are a number of challenges inherent in achieving transdisciplinarity and community participation. These have been discussed in chapter eight and only two key issues will be re-iterated here. The first relates to fostering equal partnerships among the transdisciplinary team of researchers, and between the transdisciplinary team and community members. The second relates to how concepts of *community* and *participation* are adopted and deployed.

Critical ecohealth seeks to draw attention to power dynamics that exist among professionals as they try to work together as a team. By virtue of their different positions, ranks, departments and disciplines, there is always a subtle power struggle that is not readily apparent to many. Such power struggles, if not contained, could lead to one team member dominating the research process, and overshadowing valuable contributions from other members. The principal investigator or team leader has an important role to play in ensuring that the research process equally integrates the expertise of all members of the research team. Also, as no knowledge is value-free or neutral, it is important to evaluate the forms of knowledge that are emanating from these experts, and write this into the research findings to illustrate their situated nature.

Secondly, it is usually the goal of the transdisciplinary research team to work collaboratively with local people and other stakeholders, sometimes involving them as co-researchers in investigating an environmental health problem. However, lay people and their knowledges are not as privileged as scientific understandings. So while, efforts may be made to integrate the two knowledge systems, it is important to recognize that equal integration may never be attained. Besides, it is the researcher who documents lay people's knowledge and perceptions, and such documentation may or may not accurately capture or represent local people's perspectives. Critical ecohealth requires that these shortcomings be noted and written in the text to allow for a transparent reading and interpretation of the research findings.

It is also important to distinguish between "public" and "private" accounts when working with local people (Mosse 1994). In working with a professional team of experts, most of whom are usually external researchers, local people may not be forthcoming with their issues and may put up a "public" account, compared to the real "private" account. This is especially true with participatory methodologies when participants are obliged to speak, but may not be giving "true" accounts of the issue under investigation. Mosse (1994) cautions that knowledge is not self-evident. The information generated through participatory approaches is "often of very different kinds, involving mixed combinations of fact and value, consensus and difference, openness and sensitivity, as well as public and private accounts" (p. 502). Caution has to be taken when interpreting such information, taking into consideration micro-politics and power struggles in the group and community at large.

Similarly *communities* are not pre-defined spaces, with pre-existing people. Communities should not be seen as having pre-existing features of unhealthy people or oppressed or marginalized people. Instead community members should be seen as always in the process of being constituted by various discourses (Cameron and Gibson 2005). Critical ecohealth seeks to avoid associating a particular identity with a "community" or its membership. Communities should be seen as having a membership that is embedded in a variety of social structures, with varying and conflicting needs, views, experiences and knowledges about their environments and health, and so issues have to always be contextualized.

14.9 Towards a Reflexive Ecohealth Research Practitioner

According to poststructuralists, all knowledge claims are social constructions. As social constructions, they are not value-free, but are partial, situated, and hence can be located. As a researcher informed by a particular discourse, we are all trapped in this space of creating partial knowledges. How then can we make our situatedness apparent, so as to allow the reader to better locate our views and biases in the text we write? Haraway (1991) suggests we follow a critical genealogy of subjectivity, in which we disclose our position, state our situatedness (e.g. sex, class, race) and our particular biases, so that we can be held accountable for the knowledges we produce through research, writing, reading, and interpretation.

For public health professionals and researchers, reflexive practice involves the ability critically to interrogate our use of knowledge and to examine and be cautious of the interests we serve and reproduce as we go about conducting research or practicing as professionals. Given that we cannot escape this position, it is important to be conscious of our positions as producers and reproducers of certain discourses and practices, and the values and commitments associated with the use of such discourses and practices (Fox 1991). Within the context of ecohealth research then, it is important to explicate our ideologies, and to examine the extent to which we participate in power relations surrounding the use of, adoption and circulation of environment and health discourses, and to determine whether these discourses are liberating or constraining.

Reflexive ecohealth therefore requires we pay attention to language, and various discourses, and interrogate the manner in which knowledge claims on environment, health, and development become generally accepted as the common-sense knowledge which sustains society. Reflexive ecohealth also requires we examine the ways concepts such as "health", "ecosystem degradation" "risk", "sustainability" are deployed in environment and public health research and practice.

14.10 Conclusion

This chapter has articulated a new analytical framework for ecohealth, called *critical ecohealth*. Critical ecohealth draws on recent theoretical developments in critical social theory and integrates the application of these theoretical developments in the fields of public health and environment. As an emerging field, ecohealth remains undertheorized and stands to benefit from these critical perspectives. The application of critical theory to ecohealth problems provides a thorough understanding of the causal factors and processes driving ecosystem degradation and the associated health problems. It provides a sophisticated understanding of how environmental problems and their associated health outcomes are framed. As an analytical framework, critical ecohealth cautions against the adoption of a priori notions and concepts of environment and health. It argues that the identification, explanation and resolution of environment and health problems is a political process, and must be grounded in social and political analyses.

A critical approach to ecohealth also allows for the experiences of people to be contextualized based on their multiple and changing roles, identities and experiences, so as to avoid misinterpretations and generalizations. The adoption of a critical lens to ecohealth research allows for the development of interventions that reflect biophysical realities, while being socially and politically conscious. The application of critical theoretical approaches ensures that concepts, such as transdisciplinarity, community, participation, gender, and indigenous knowledge systems are not adopted and used uncritically, as this may not only fail to yield effective interventions, but could augment inequities surrounding environment and health issues.

References

Abu-Lughod L (1993) Introduction. In: Abu-Lughod L (ed) Writing women's world. Bedouin stories. University of California Press, Berkeley, pp 1–42

Adger WN, Benjaminsen TA, Brown K, Svarstad H (2001) Advancing a political ecology of global environmental discourses. Dev Change 32:681–715

Annandale E, Clark J (1996) What is gender? Feminist theory and the sociology of reproduction. Sociol Health Illn 18(1):17–44

Barnes B (1982) T.S. Kuhn and social science. Macmillan, London

Blaikie P (1995) Understanding environmental issues. In: Morse S, Stocking M (eds) People and environment. University College London Press, London, pp 1–30

Bryant RL (1997) Third world political ecology: an introduction. In: Bryant RL and Bailey S(ed) Third world political ecology. Routledge, London, New York, NY

Bryant RL, Bailey S (1997) Third world political ecology. Routledge, London

Butler J (1990) Gender trouble. Routledge, London

Butler J (1992) Contingent foundations: feminism and the question of "postmodernism". In: Butler J and Scott JW (eds) Feminists theorize the political. Routledge, New York, NY, pp 3–21

Cameron J, Gibson K (2005) Alternative pathways to community and economic development: The Latrobe Valley community partnering project. Geogr Res 43(3):274–285

Dakubo C (2004) Ecosystem approach to community health planning in Ghana. EcoHealth 1: 50–59

Escobar A (1996) Constructing nature: elements for a poststructural political ecology. In: Peet R, Watts M (eds) Liberation ecologies: environment, development, and social movements. Routledge, London, pp 46–68

Fox NJ (1991) Postmodernism, rationality and the evaluation of health care. Sociol Rev 39(4): 709–744

Fox NJ (1994) Postmodernism, sociology and health. University of Toronto Press, Toronto, ON

Freundenberg WR, Frickel S, Gramling R (1995) Beyond the nature/society divide: learning to think about a mountain. Sociol Forum 10:361–392

Frug MJ (1992) Postmodern legal feminism. Routledge, London

Guthman J (1997) Representing crisis: the theory of Himalayan environmental degradation and the project of development in post-rana Nepal. Dev Change 28:45–69

Haraway D (1991) Simians, cyborgs and women: the reinvention of nature. Free Association, London

Mackenzie F (1995) Selective silence: a feminist encounter with the environmental discourse in colonial Africa. In: Crush J (ed) Power of development. Routledge, New York, NY, pp 100–112, 324 pp

Mosse D (1994) Authority, gender and knowledge: theoretical reflections on the practice of participatory rural appraisal. Dev Change 25(2):497–526

Rocheleau DE (1995) Gender and biodiversity: a feminist political ecology perspective. IDS Bull 26:9–16

Schmink M, Wood CH (1987) The 'political ecology' of Amazonia. In: Little PD and Horowitz MM, (eds) Lands at risk in the third world: local-level perspectives. Westview Press, Boulder, CO, pp 38–57

Smart C (1990) Feminist approaches to criminology or postmodern woman meets atavistic man. In: Gelsthorpe L, Morris A (eds) Feminist perspectives in criminology. Open University Press, Milton Keynes

Spivak G, Ronney E (1989) 'In a word' interview. Differences 1:124–156

Whiteford LM, Manderson L (2000) Global health policy, local realities. The fallacy of the level playing ground. Lynne Rienner Publishers, Boulder London

Whittle KS, Inhorn MC (2001) Rethinking difference: a feminist reframing of gender/race/class for improvement of women's health research. Int J Health Serv 31(1):147–165

Index

Note: The letters 'f' and 't' followed by the locators refers to figures and tables respectively.

C.Y. Dakubo, *Ecosystems and Human Health*, DOI 10.1007/978-1-4419-0206-1,
© Springer Science+Business Media, LLC 2011

LaVergne, TN USA
06 December 2010
207592LV00002B/88/P